FIELD GUIDE TO THE BIRDS OF SOUTHERN AFRICA

To my parents

IAN SINCLAIR'S

FIELD GUIDE TO THE BIRDS OF SOUTHERN AFRICA

THE STEPHEN GREENE PRESS
Lexington, Massachusetts

Cover photographs: (top left) Woodland Kingfisher (J. van Jaarsveld); (top right) Whitefaced Owls (J. van Jaarsveld); (bottom left) Angola Pitta (D.C.H. Plowes); (bottom right) Black Harrier (R. Jones).

First published in South Africa by C. Struik Publishers, 1984
First published in the United States of America by
The Stephen Greene Press, Inc., 1987
Distributed by Viking Penguin Inc., 40 West 23rd Street,
New York, NY 10010

Distribution maps reproduced from Kenneth Newman's *Newman's Birds of Southern Africa* with kind permission of the author and the publisher, Macmillan South Africa.

ISBN 0-8289-0621-1

CIP data available

Printed in South Africa
by National Book Printers, Goodwood

set in 8 on 9 pt Helvetica Light

CONTENTS

FOREWORD

In the space of a short foreword I can do little more than briefly introduce the author and the subject of his book. 'Ian' Sinclair (only a few close friends are privy to the names represented by his initials J.C.) became a member of the staff of the FitzPatrick Institute of African Ornithology at the University of Cape Town in 1976. Six years later he left the Institute to get married and to take up his present post in Durban.

This book had its genesis in 1979 while Ian was a member of the FitzPatrick Institute's team of researchers on Marion Island. Much of the text was first drafted during his stay on the island. I believe that Ian's association with the FitzPatrick Institute helped him to develop the discipline necessary for translating his exceptional abilities for finding and identifying birds in the field into something more durable and of benefit to others.

The prototype for modern field guides was developed in the 1930s. In those days a tweed sports-coat was *de rigueur* for gentlemen bird-watchers. Hence, the size of a typical field guide came to be set by the capacity of the pocket of a standard sports-coat. Accommodating southern Africa's avifauna of some 900 species within a pocket-size book presents problems, and particularly so in this case in which photographs, rather than paintings, are used to illustrate the birds. Consequently, economies had to be made and, for example, the non-breeding plumages of some species are not illustrated. This is compensated for by the fact that the photographs chosen are those that most clearly show the important field characters of each species, and the guide will therefore be a significant aid to field identification, especially for novice birders.

Ian's guide is a welcome addition to the ornithological literature of southern Africa and I look forward to the Afrikaans edition planned for the early part of next year. There is a crying need for Afrikaans field guides dealing with the flora and fauna of southern Africa. A greater public awareness of the sub-continent's rich natural history is part of an advance towards realization of an optimal use of all the region's natural resources.

PROFESSOR W.R. SIEGFRIED
DIRECTOR, PERCY FITZPATRICK INSTITUTE OF AFRICAN ORNITHOLOGY
UNIVERSITY OF CAPE TOWN

ACKNOWLEDGEMENTS

The draft text of this book was prepared in 1979, during the long wintery nights on Marion Island in the sub-Antarctic where I was researching seabirds for the Percy FitzPatrick Institute of African Ornithology. My thanks go to the Institute's Director, Professor Roy Siegfried, who gave me both the opportunity and the incentive to produce the draft text. While I was at the Institute, many members of the staff gave of their time in helpful discussion and in sharing information on problematic bird identifications but, in particular, I am indebted to Rodney Cassidy, Phil Hockey, Richard Brooke, Alan Burger, Tamar Salinger, John Cooper and Andy Griffiths. This book must be seen as one of the many contributions that the FitzPatrick Institute has made to bird-watchers in this country.

On my return from Marion Island, Alan and Meg Kemp took my rambling draft text and, between them, condensed and retyped the manuscript. To them both go my thanks for their encouragement and interest in the project.

Gerry Nicholls has been a great source of inspiration and enthusiasm from the inception of this book, and he assisted in preparing the final text on waders, gulls and terns.

I am also grateful to my colleague and partner in the 'twitching' game, Aldo Berruti, for reading the manuscript in its final form and for his many helpful and constructive comments.

The many photographers who scoured their collections to supply transparencies for this book gained both my great appreciation and admiration: the following pages reflect the superb standard of bird photography in southern Africa. Alan Wilson gave unstintingly of his time to pursue previously unphotographed birds especially for this book, and I thank him for his patience with my requests.

Artist Peter Harrison, based in England, painted the fine illustrations of the species for which no photographs were available. Many of the species are little-known birds that few people have ever seen, and Peter is to be congratulated for catching their 'jizz' and portraying them as they would appear in the field.

I am indebted to Ken Newman and his publisher, Macmillan South Africa, for their permission to use the basic distribution maps from **Newman's Birds of Southern Africa.** Richard Brooke amended and updated them and to him go my most sincere thanks for executing this task.

Finally, Eve Gracie, the editor of this book who also undertook successfully to gather the thousands of photographs needed for the final selection, has made work on this project a pleasant and memorable experience and her own enthusiasm for the subject kept me on my toes in the final stages.

IAN SINCLAIR
MAY 1984

INTRODUCTION

The geographical region covered in this book is that of southern Africa south of the Cunene and Zambezi rivers, the southern oceans due south of Africa to Antarctica and the relevant sub-Antarctic and Antarctic islands. The text covers a total of 913 species recorded within this region, of which over 140 are endemic or near endemic species. (Near endemics are those which have their major populations within the region and just marginally spill over or are vagrant to neighbouring territories, such as Angola and Zambia.) All species, including vagrants, found within the region are treated equally in the text, and are illustrated in full colour.

This field guide is not intended as a replacement of a book with paintings, where an artist is able to illustrate sexual dimorphism and the different age groups of species. Instead, it uses photographs of the birds; photographs that have been specifically chosen to show the most important field characters. In many instances the males are depicted because they frequently show the more obvious plumage coloration and are therefore more easily identified. Equally, many birds, such as seabirds and raptors, are shown in flight because that is how they are most usually seen by the birder.

Photographs of birds at their nests have been avoided wherever possible. The reason for this is because most birders never see birds in this domestic situation, besides which, photographing birds at the nest is a deliberate disturbance and a cause of desertion, and something not to be encouraged. In some overseas countries it is illegal to photograph certain categories of birds at their nests.

Where an illustration has been used to depict a bird, it was because no photograph was available, and I urge all bird photographers to get into the field and obtain photographs of these species for inclusion in future editions of this book. If you think you have a better photograph than the one appearing in the book, please let me know about it.

In the few instances where neither a photograph nor an illustration has been used, this is because the validity of the bird's occurrence in the region is in doubt.

The text prepared for this book gives the information necessary to enable each individual species to be identified when compared with those it closely resembles. Those characters irrelevant to the bird's identification have been omitted. In some instances, the taxonomic grouping of species has been juxtaposed to enable direct comparison with similar-looking birds. Many of the identification criteria presented are original and the result of a number of years spent in the field, studying bird identification in southern Africa.

HOW TO USE THIS BOOK

Each species' account is prefaced by the number allocated to it by the Southern African Ornithological Society. Following the common and scientific names, the length of the bird is given: this indicates not the bird's height but a measurement from the bill tip to the tail tip of an outstretched bird. In those cases where a bird has an exceptionally long tail, this is given as a separate measurement; otherwise, for example, the overall length of a Pintailed Whydah would be equal to that of a Feral Pigeon which is, in reality, ten times its bulk.

BIRD TOPOGRAPHY

Different parts of a bird's external anatomy are referred to throughout this book. This is not highbrow birding language but necessary terminology to indicate the exact position of identifying characters. When confronted with an unknown species, a quick and accurate description should be taken to enable a later positive identification to be made from a reference work. Study the following sketches of bird topography and become familiar with the different parts, as this knowledge will prove invaluable at a later stage when facing complex identifications.

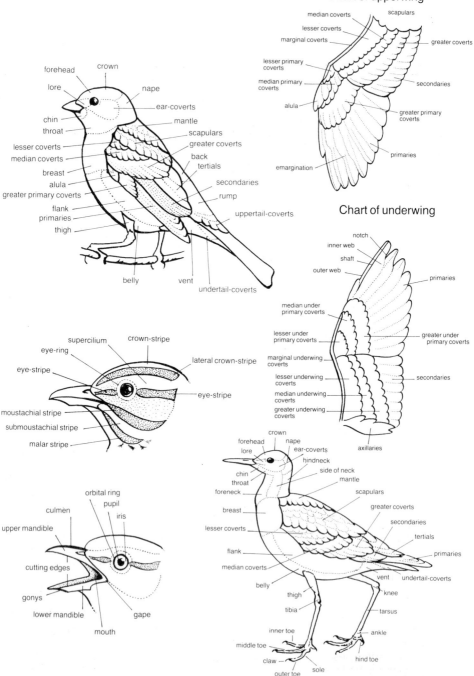

Chart of upperwing

scapulars
median coverts
lesser coverts
marginal coverts
greater coverts
lesser primary coverts
median primary coverts
secondaries
alula
greater primary coverts
emargination
primaries

forehead
crown
lore
nape
ear-coverts
chin
mantle
throat
scapulars
lesser coverts
greater coverts
median coverts
back
breast
tertials
alula
secondaries
greater primary coverts
rump
flank
uppertail-coverts
primaries
thigh
belly
vent
undertail-coverts

Chart of underwing

notch
inner web
shaft
outer web
primaries
median under primary coverts
lesser under primary coverts
greater under primary coverts
marginal underwing coverts
lesser underwing coverts
secondaries
median underwing coverts
greater underwing coverts
axillaries

supercilium
crown-stripe
eye-ring
lateral crown-stripe
eye-stripe
eye-stripe
moustachial stripe
submoustachial stripe
malar stripe

orbital ring
pupil
culmen
iris
upper mandible
cutting edges
gonys
lower mandible
gape
mouth

crown
forehead
nape
lore
ear-coverts
hindneck
chin
side of neck
throat
mantle
foreneck
scapulars
breast
greater coverts
lesser coverts
secondaries
tertials
flank
primaries
median coverts
vent
belly
undertail-coverts
thigh
knee
tibia
tarsus
inner toe
ankle
middle toe
claw
hind toe
outer toe
sole

HABITAT

An important clue to the bird's identity is its habitat. For instance, it is highly unlikely that you will see a forest robin foraging alongside waders on a mudflat. Most species have a favoured habitat, especially when breeding, but they often wander and when migrating they may travel through many different habitats.

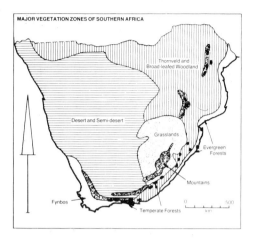

The map shows the major vegetation zones of the region. Bird distribution is contained by, and in many cases, restricted to specialized habitats within these zones. The habitats can be divided as follows:

Cities, towns and gardens. Although at a first glance appearing virtually 'dead' areas for birdlife, any of our major cities and their suburbs have lists in excess of a hundred species. Falcons, hawks, robins, cuckoos, warblers, weavers, and the ubiquitous doves and sparrows, all occur regularly in any urban habitat. Many species can thrive in well-vegetated gardens, especially those featuring indigenous plants. More species can be attracted by installing bird feeders and tables and, in the drier regions, by supplying a regular source of water. Being a city-dweller should be no hindrance to enjoying bird-watching. Visit your local botanical garden, open parks, nature reserves or sewage farm and you will be surprised at the diversity of birds they harbour.

Grasslands and cultivated lands. The major grassland areas exist in the Transvaal, Orange Free State and Natal, although pockets are found throughout southern Africa. The more common bird families associated with this habitat are bustards, cranes, larks, pipits, cisticolas, canaries, widowbirds and francolins. Cultivated lands, depending on the crop planted, tend to attract a great abundance of a few species. However, sugar cane, exotic pine and eucalyptus plantations, cultivated at the expense of natural vegetation, are known for their dearth of birdlife. The basic lack of natural food in such plantations makes them inhospitable to most species although a few have adapted to some degree in various regions.

Wetlands. Encompassing the smallest pond or stream to the largest natural lake, swamp or man-made dam. Certain families, such as ducks, cormorants, flamingoes and pelicans occur only on water while other families, for example some ibises, egrets and herons, feed far from water but either nest near it or roost on it or in surrounding vegetation. Reedbeds harbour shy, skulking birds, such as bitterns, crakes and flufftails; interesting species, but

11

often impossible to see unless they are flushed. Lake St Lucia, the Okavango swamps and the Nylsvlei floodplain are the finest examples of this habitat.

Open sea and coast. Includes estuaries like Walvis Bay and saline lagoons such as those found at Richard's Bay. Muddy estuaries invariably attract flocks of gulls, terns, and waders, sometimes in their thousands, from the northern hemisphere. In comparison, except over large stands of seaweed and rocky headlands, the open coastline has less diversity of bird species. The abundance of seabirds over the open ocean, however, is generally underestimated as the albatrosses, petrels and shearwaters are not readily seen from shore, other than from headlands during inclement weather. A berth on a deep-sea fishing boat or trawler provides the best opportunity to see these birds and the further out to sea the vessel travels, the better. Good ports to travel from during summer and winter are Hout Bay and Saldanha Bay, while the best winter-only ports are Port Elizabeth, East London and Durban.

Thornveld and broad-leafed woodland. This habitat harbours the greatest diversity of bird species in the region and most can be easily observed. Acacia, miombo and mopane trees comprise the major part of this 'bushveld' which is most evident in the northern part of southern Africa. Many species such as the Spotted Creeper, Boulder Chat, Cabanis's Bunting, Miombo Rock Thrush and Mashona Hyliota are peculiar to miombo (*Brachystegia*) woodland. Thornveld is obvious in the game reserves of northern Zululand, and the Kruger National Park has both thornveld and mopane woodland. Large tracts of miombo woodland are found in Zimbabwe and Moçambique.

Mountains. Here we consider the mountains of the Cape and the higher ranges in Natal, the Transvaal and Zimbabwe. At higher elevations, above the tree and forest line, the number of bird species found is low but those that do occur are interesting: the rockjumpers, pipits, chats and siskins. Stands of flowering proteas and aloes attract many species, such as the endemic sugarbirds and Protea Canary, and hosts of sunbirds. However, the richest birding areas are not in the higher reaches but are the well-wooded kloofs and grassy and bracken-covered valleys, where large numbers of birds congregate, especially in winter when many descend from the higher elevations. Sir Lowry's Pass, Bain's Kloof and Piekenier's Pass in the Cape are all superb birding areas. The Drakensberg range is very good throughout, especially above an altitude of 1 500 metres.

Desert and semi-desert. This is the vast, low-rainfall area of the west and south-west and that in which the majority of birds endemic to southern Africa occur: Karoo and Rüppell's korhaans, the Namaqua Sandgrouse and Prinia, Karoo Chat and Lark, Tractrac and Sicklewinged chats and most of the lark species.
   The vegetation is sparse and large numbers of birds are nomadic within the area, following rain fronts when they occur. Good birding areas include the Namib Naukluft Park in the Namib Desert, which offers a variety of desert habitats, and the northern Cape just south of the Orange River between Springbok and Kimberley. Virtually anywhere in the Karoo, particularly after rain, can provide excellent endemic birding.

Evergreen forests. Along the eastern Cape coast north to Moçambique, and in pockets in the mountainous terrain of Natal, the Transvaal and Zimbabwe, are dense evergreen forests with a thick understorey. These superb areas should not be confused with birdless exotic plantations, as evergreen forests are generally situated in areas of high rainfall and mist belts, and are extremely rich in birdlife. However, the robins, thrushes, warblers and shrikes occurring within them are not conspicuous, tending to skulk and remain furtive in the leafy canopy or thicker tangles. Outstanding coastal forests are Dwesa in the Transkei, and Mapelane and Cape Vidal in Zululand. Inland forests that provide rewarding birding are the Karkloof range in Natal, Woodbush near Tzaneen in the Transvaal, and the Vumba and Inyanga forests in Zimbabwe.

MAPS AND BIRD DISTRIBUTION

Bird distribution is in a continual state of flux and, although usually on a small scale, whole populations have been known to move, often influenced by climatic cycles.

A map is included to show, as accurately as is possible within this format, each species' current distribution throughout the region. On a map of this size exact distribution is not feasible because the region covered is too vast; for precise distribution it is best to consult bird atlases for smaller regions or provinces.

Apart from indicating where a species occurs, the distribution map also helps towards a bird's identification in that it immediately shows if that species is found in your region. For instance, a small gull seen on the Natal coast is more likely to be a Greyheaded because the similar Hartlaub's is virtually unknown there, being confined to the west coast.

In the text, unless otherwise stated, a bird is taken to be resident within its distribution range, or else it undertakes local migrations within the boundaries of the region.

SOUTHERN AFRICAN PLACE NAMES AND LOCATIONS

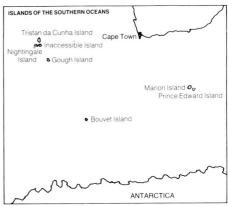

ISLANDS OF THE SOUTHERN OCEANS

BIRD VOCALS

These can be divided into two categories: the song, which is given to proclaim a territory, and the call, which is used in alarm or contact. The song, usually uttered during the breeding season, is loud and consists of a prolonged series of notes, whereas the call is normally short and abrupt.

The call of each bird is described if it has been recorded in the region, but the difficulty of achieving an accurate, written rendition of a bird's song or call is well known. Words cannot satisfactorily evoke songs or calls, and graphs of bird vocals, known as sonograms, still leave most people none the wiser as to the actual sound. The importance of vocals in bird identification cannot be over-stressed and indeed, many species, especially the cisticolas and some pipits, are indistinguishable in the field unless their call is heard.

Many records and tapes of bird songs and calls are available and these are invaluable in learning to identify calls. However, their greatest use is in the field, as many bird species will respond readily to a tape of their own call or song and will approach to investigate the source. In the field, a small, hand-held tape recorder and a selection of tapes are almost as indispensable as a pair of binoculars.

FIELD EQUIPMENT

A notebook, pencil and a pair of binoculars are the basic tools of a birder. The notebook and pencil can be replaced by a compact tape recorder on which to record details of observations, but binoculars cannot be dispensed with. A good scope is also invaluable.

Binoculars. A bewildering range of binoculars is available and to make an initial choice is difficult. Advertisements highlighting the super-powerful binoculars with magnifications of 15 and 20 power can be misleading: the binoculars are often of very poor quality, are heavy and totally unwieldy, and will give you eyestrain if used for prolonged periods.

It has been proven that the best combination for bird-watching is 8, 9 or 10 power with either a 30, 40 or 50 mm object lens. Check the minimal focussing distance as this can be important in close-up forest work. There are widely divergent opinions as to the make of binoculars to use but, as with most merchandise, the higher the price, the better the quality. The most expensive and best quality binoculars available are the Leitz range of roof-prism field glasses. These are the 'Rolls-Royce' of binoculars and I highly recommend them, especially the 10 x 40 combination which is ideal for birding in African conditions.

Scopes. Although often called telescopes, most of today's models do not have draw tubes, working instead by means of prisms. A scope is essential for bird-watching on estuaries, in open veld and on the coast. Birds in these habitats are frequently seen at great distances and are mostly unapproachable, but using a scope one can 'zoom' in and identify the species with ease. Again, a wide range of scopes is available and my advice is not to purchase those used for star-gazing as they are far too powerful for basic field work and are also normally too cumbersome to carry around. A well-proven and ideal scope for birding is the Kowa TS 1 with a wide-angle 20x lens. When fitted to a tripod this scope is probably as essential to the serious birder as a pair of binoculars.

Tape recorders. When selecting a small, hand-held recorder, choose a model with good volume output. Also make sure that it is comfortable to hold and can fit into a pocket. Avoid the type that has mini-tapes because these are only useful for note-taking and not for playing or recording bird calls. Check that the quality of sound output at the highest volume is good and that no deafening rumbling or hissing sounds override the bird's call. A small directional microphone is very useful for recording a bird's call or song and the model available is the Tect UEM-83. The condenser microphone that is built into most small tape recorders is unsuitable for taping bird calls.

ABBREVIATIONS AND SYMBOLS

The following abbreviations and symbols have been used in this book:

Text
ad. = adult
ads. = adults
imm. = immature
imms. = immatures
L = length
H = height

cm = centimetre
m = metre(s)
S = South
N = North
♂ = male
♀ = female
□ = not illustrated

Maps
○ = species recorded as vagrant

CHECKLIST OF SOUTHERN AFRICAN BIRDS: Numbers and names

Revised bird names and numbers together with the old 'Roberts' numbers. The first column gives the new number, the second column the old 'Roberts' number, and the third column the new name followed by any alternative name in brackets.

1	1	Ostrich	51	45	Masked Booby
2	-	King Penguin	52	46	Brown Booby
3	2	Jackass Penguin	53	44	Cape Gannet
4	3	Rockhopper Penguin	54	–	Australian Gannet
5	—	Macaroni Penguin	55	47	Whitebreasted Cormorant
6	4	Great Crested Grebe	56	48	Cape Cormorant
7	5	Blacknecked Grebe	57	49	Bank Cormorant
8	6	Dabchick (Little Grebe)	58	50	Reed Cormorant
9	—	Royal Albatross	59	51	Crowned Cormorant
10	7	Wandering Albatross	60	52	Darter
11	11	Shy Albatross	61	53	Greater Frigatebird
12	8	Blackbrowed Albatross	62	54	Grey Heron
13	9	Greyheaded Albatross	63	55	Blackheaded Heron
14	10	Yellownosed Albatross	64	56	Goliath Heron
15	12	Darkmantled Sooty Albatross	65	57	Purple Heron
16	12X	Lightmantled Sooty Albatross	66	58	Great White Egret
17	13	Southern Giant Petrel	67	59	Little Egret
18	–	Northern Giant Petrel	68	60	Yellowbilled Egret
19	15	Antarctic Fulmar (Silvergrey Petrel)	69	64	Black Egret
			70	64X	Slaty Egret
20	15X	Antarctic Petrel	71	61	Cattle Egret
21	14	Pintado Petrel	72	62	Squacco Heron
22	–	Bulwer's Petrel	73	62X	Madagascar Squacco Heron
23	16	Greatwinged Petrel	74	63	Greenbacked Heron
24	19	Softplumaged Petrel	75	65	Rufousbellied Heron
25	–	Whiteheaded Petrel	76	69	Blackcrowned Night Heron
26	18	Atlantic Petrel (Schlegel's Petrel)	77	70	Whitebacked Night Heron
			78	67	Little Bittern
27	–	Kerguelen Petrel	79	66	Dwarf Bittern (Rail Heron)
28	20	Blue Petrel	80	71	Bittern
29	21, 21X, 22	Broadbilled Prion	81	72	Hamerkop
			82	–	Shoebill
30	22Y	Slenderbilled Prion	83	80	White Stork
31	22X	Fairy Prion	84	79	Black Stork
32	23	Whitechinned Petrel	85	78	Abdim's Stork (Whitebellied Stork)
33	24	Grey Shearwater			
34	26	Cory's Shearwater	86	77	Woollynecked Stork
35	25	Great Shearwater	87	74	Openbilled Stork (Openbill)
36	26X	Fleshfooted Shearwater			
37	29	Sooty Shearwater	88	75	Saddlebilled Stork (Saddlebill)
38	26Y	Manx Shearwater			
39	27	Little Shearwater	89	73	Marabou Stork
40	28	Audubon's Shearwater	90	76	Yellowbilled Stork
41	–	Wedgetailed Shearwater	91	81	Sacred Ibis
42	30	European Storm Petrel	92	82	Bald Ibis
43	31	Leach's Storm Petrel	93	83	Glossy Ibis
44	33	Wilson's Storm Petrel	94	84	Hadeda Ibis
45	37	Whitebellied Storm Petrel	95	85	African Spoonbill
46	36	Blackbellied Storm Petrel	96	86	Greater Flamingo
47	39	Redtailed Tropicbird	97	87	Lesser Flamingo
48	40	Whitetailed Tropicbird	98	87X	Mute Swan
49	42	White Pelican	99	100	Whitefaced Duck
50	41	Pinkbacked Pelican	100	101	Fulvous Duck

101	104	Whitebacked Duck	162	165	Pale Chanting Goshawk
102	89	Egyptian Goose	163	163	Dark Chanting Goshawk
103	90	South African Shelduck	164	—	European Marsh Harrier
104	96	Yellowbilled Duck	165	167	African Marsh Harrier
105	95	African Black Duck	166	170	Montagu's Harrier
106	98	Cape Teal	167	168	Pallid Harrier
107	99	Hottentot Teal	168	169	Black Harrier
108	97	Redbilled Teal	169	171	Gymnogene
109	99X	Pintail	170	172	Osprey
110	97X	Garganey	171	113	Peregrine Falcon
111	93	European Shoveler	172	114	Lanner Falcon
112	94	Cape Shoveler	173	115	Hobby Falcon
113	102	Southern Pochard	174	116	African Hobby Falcon
114	92	Pygmy Goose	175	116Y	Sooty Falcon
115	91	Knobbilled Duck	176	116X	Taita Falcon
116	88	Spurwinged Goose	177	—	Eleonora's Falcon
117	103	Maccoa Duck	178	117	Rednecked Falcon
118	105	Secretarybird	179	120	Western Redfooted Kestrel
119	150	Bearded Vulture	180	119	Eastern Redfooted Kestrel
		(Lammergeier)	181	123	Rock Kestrel
120	111	Egyptian Vulture			(Common Kestrel)
121	110	Hooded Vulture	182	122	Greater Kestrel
122	106	Cape Vulture	183	125	Lesser Kestrel
123	107	Whitebacked Vulture	184	121X	Grey Kestrel
124	108	Lappetfaced Vulture	185	121	Dickinson's Kestrel
125	109	Whiteheaded Vulture	186	126	Pygmy Falcon
126	128	Black Kite	187	188X	Chukar Partridge
	129	Yellowbilled Kite	188	173	Coqui Francolin
127	130	Blackshouldered Kite	189	174	Crested Francolin
128	127	Cuckoo Hawk	190	176	Greywing Francolin
129	131	Bat Hawk	191	177	Shelley's Francolin
130	132	Honey Buzzard	192	178	Redwing Francolin
131	133	Black Eagle	193	179	Orange River Francolin
		(Verreaux's Eagle)	194	182	Redbilled Francolin
132	134	Tawny Eagle	195	181	Cape Francolin
133	135	Steppe Eagle	196	183	Natal Francolin
134	136	Lesser Spotted Eagle	197	184	Hartlaub's Francolin
135	137	Wahlberg's Eagle	198	188	Rednecked Francolin
136	139	Booted Eagle	199	185	Swainson's Francolin
137	141	African Hawk Eagle	200	189	Common Quail
138	140	Ayres' Eagle	201	190	Harlequin Quail
139	138	Longcrested Eagle	202	191	Blue Quail
140	142	Martial Eagle	203	192	Helmeted Guineafowl
141	143	Crowned Eagle	204	193	Crested Guineafowl
142	145	Brown Snake Eagle	205	196	Kurrichane Buttonquail
143	146	Blackbreasted Snake Eagle	206	194	Blackrumped Buttonquail
		(Short-toed Eagle)			(Hottentot Buttonquail)
144	147	Southern Banded Snake Eagle	207	215	Wattled Crane
145	148	Western Banded Snake Eagle	208	216	Blue Crane
146	151	Bateleur	209	214	Crowned Crane
147	112	Palmnut Vulture	210	197	African Rail
148	149	African Fish Eagle	211	198	Corncrake
149	154	Steppe Buzzard	212	199	African Crake
150	155	Forest Buzzard	213	203	Black Crake
		(Mountain Buzzard)	214	201	Spotted Crake
151	—	Longlegged Buzzard	215	202	Baillon's Crake
152	152	Jackal Buzzard	216	200	Striped Crake
153	152B	Augur Buzzard	217	205	Redchested Flufftail
154	144	Lizard Buzzard	218	206	Buffspotted Flufftail
155	156	Redbreasted Sparrowhawk	219	207X	Streakybreasted Flufftail
156	157	Ovambo Sparrowhawk	220	206X	Longtoed Flufftail
157	158	Little Sparrowhawk	221	207	Striped Flufftail
158	159	Black Sparrowhawk	222	204	Whitewinged Flufftail
159	161	Little Banded Goshawk (Shikra)	223	208	Purple Gallinule
160	160	African Goshawk	224	209	Lesser Gallinule
161	162	Gabar Goshawk	225	208X	American Purple Gallinule

16

226	210	Moorhen
227	211	Lesser Moorhen
228	212	Redknobbed Coot
229	213	African Finfoot
230	217	Kori Bustard
231	219	Stanley's Bustard
		(Denham's Bustard)
232	218	Ludwig's Bustard
233	222	Whitebellied Korhaan
234	223	Blue Korhaan
235	220	Karoo Korhaan
236	221	Rüppell's Korhaan
237	224	Redcrested Korhaan
238	227	Blackbellied Korhaan
239	225	Black Korhaan
240	228	African Jacana
241	229	Lesser Jacana
242	230	Painted Snipe
243	231X	European Oystercatcher
244	231	African Black Oystercatcher
245	233	Ringed Plover
246	235	Whitefronted Plover
247	236	Chestnutbanded Plover
248	237	Kittlitz's Plover
249	238	Threebanded Plover
250	234	Mongolian Plover
251	239	Sand Plover
252	240	Caspian Plover
253	240X	Lesser Golden Plover
254	241	Grey Plover
255	242	Crowned Plover
256	244	Lesser Blackwinged Plover
257	243	Blackwinged Plover
258	245	Blacksmith Plover
259	246	Whitecrowned Plover
260	247	Wattled Plover
261	248	Longtoed Plover
		(Whitewinged Plover)
262	232	Turnstone
263	257	Terek Sandpiper
264	258	Common Sandpiper
265	259	Green Sandpiper
266	264	Wood Sandpiper
267	–	Spotted Redshank
268	261	Redshank
269	262	Marsh Sandpiper
270	263	Greenshank
271	254	Knot
272	251	Curlew Sandpiper
273	251Y	Dunlin
274	253	Little Stint
275	–	Longtoed Stint
276	253X	Rednecked Stint
277	–	Whiterumped Sandpiper
278	252	Baird's Sandpiper
279	251X	Pectoral Sandpiper
280	–	Temminck's Stint
281	255	Sanderling
282	–	Buffbreasted Sandpiper
283	251Z	Broadbilled Sandpiper
284	256	Ruff
285	249	Great Snipe
286	250	Ethiopian Snipe
287	265	Blacktailed Godwit
288	266	Bartailed Godwit
289	267	Curlew
290	268	Whimbrel
291	271	Grey Phalarope
292	272	Rednecked Phalarope
293	–	Wilson's Phalarope
294	269	Avocet
295	270	Blackwinged Stilt
296	273	Crab Plover
297	275	Spotted Dikkop
		(Cape Dikkop)
298	274	Water Dikkop
299	276	Burchell's Courser
300	277	Temminck's Courser
301	278	Doublebanded Courser
302	279	Threebanded Courser
303	280	Bronzewinged Courser
304	281	Redwinged Pratincole
		(Collared Pratincole)
305	282	Blackwinged Pratincole
306	283	Rock Pratincole
		(Whitecollared Pratincole)
307	284	Arctic Skua
308	–	Longtailed Skua
309	285	Pomarine Skua
310	286	Subantarctic Skua
311	–	South Polar Skua
312	287	Kelp Gull
		(Southern Blackbacked Gull)
313	287X	Lesser Blackbacked Gull
314	–	Herring Gull
315	288	Greyheaded Gull
316	289	Hartlaub's Gull
317	–	Franklin's Gull
318	289X	Sabine's Gull
319	–	Blackheaded Gull
320	–	Blacklegged Kittiwake
321	290X	Gullbilled Tern
322	290	Caspian Tern
323	–	Royal Tern
324	298	Swift Tern
		(Greater Crested Tern)
325	297	Lesser Crested Tern
326	296	Sandwich Tern
327	291	Common Tern
328	294	Arctic Tern
329	292	Antarctic Tern
330	293	Roseate Tern
331	–	Blacknaped Tern
332	295	Sooty Tern
333	295X	Bridled Tern
334	300	Damara Tern
335	299	Little Tern
336	–	Whitecheeked Tern
337	305X	Black Tern
338	305	Whiskered Tern
339	304	Whitewinged Tern
340	303	Common Noddy
341	–	Lesser Noddy
342	302	Fairy Tern
343	306	African Skimmer
344	307	Namaqua Sandgrouse
345	308	Burchell's Sandgrouse
		(Spotted Sandgrouse)
346	309	Yellowthroated Sandgrouse
347	310	Doublebanded Sandgrouse
348	–	Feral Pigeon
349	311	Rock Pigeon
		(Speckled Pigeon)

350	312	Rameron Pigeon	413	381	Bradfield's Swift
351	313	Delegorgue's Pigeon	414	379	Pallid Swift
		(Bronzenaped Pigeon)	415	383	Whiterumped Swift
352	314	Redeyed Dove	416	384	Horus Swift
353	315	Mourning Dove	417	385	Little Swift
354	316	Cape Turtle Dove	418	386	Alpine Swift
355	317	Laughing Dove	419	382	Mottled Swift
356	318	Namaqua Dove	420	385X	Scarce Swift
357	320	Bluespotted Dove	421	387	Palm Swift
358	321	Greenspotted Dove	422	388	Mottled Spinetail
		(Emeraldspotted Dove)	423	389	Böhm's Spinetail
359	319	Tambourine Dove			(Batlike Spinetail)
360	322	Cinnamon Dove	424	390	Speckled Mousebird
361	323	Green Pigeon	425	391	Whitebacked Mousebird
362	326	Cape Parrot	426	392	Redfaced Mousebird
		(Brownnecked Parrot)	427	393	Narina Trogon
363	328	Brownheaded Parrot	428	394	Pied Kingfisher
364	327	Meyer's Parrot	429	395	Giant Kingfisher
365	329	Rüppell's Parrot	430	396	Halfcollared Kingfisher
366	329X	Roseringed Parakeet	431	397	Malachite Kingfisher
367	330	Rosyfaced Lovebird	432	398	Pygmy Kingfisher
368	332	Lilian's Lovebird	433	399	Woodland Kingfisher
369	331	Blackcheeked Lovebird	434	400	Mangrove Kingfisher
370	336	Knysna Lourie	435	402	Brownhooded Kingfisher
371	337	Purplecrested Lourie	436	401	Greyhooded Kingfisher
372	–	Ross's Lourie	437	403	Striped Kingfisher
373	339	Grey Lourie	438	404	European Bee-eater
374	340	European Cuckoo	439	406	Olive Bee-eater
375	341	African Cuckoo	440	405	Bluecheeked Bee-eater
376	342	Lesser Cuckoo	441	407	Carmine Bee-eater
377	343	Redchested Cuckoo	442	408	Böhm's Bee-eater
378	344	Black Cuckoo	443	409	Whitefronted Bee-eater
379	344X	Barred Cuckoo	444	410	Little Bee-eater
380	346	Great Spotted Cuckoo	445	411	Swallowtailed Bee-eater
381	347	Striped Cuckoo	446	412	European Roller
382	348	Jacobin Cuckoo	447	413	Lilacbreasted Roller
383	345	Thickbilled Cuckoo	448	414	Racket-tailed Roller
384	350	Emerald Cuckoo	449	415	Purple Roller
385	351	Klaas's Cuckoo	450	416	Broadbilled Roller
386	352	Diederik Cuckoo	451	418	Hoopoe
387	358	Green Coucal	452	419	Redbilled Woodhoopoe
388	353	Black Coucal	453	420	Violet Woodhoopoe
389	354	Copperytailed Coucal	454	421	Scimitarbilled Woodhoopoe
390	355	Senegal Coucal			(Scimitarbill)
391	356	Burchell's (Whitebrowed) Coucal	455	422	Trumpeter Hornbill
392	359	Barn Owl	456	423	Silverycheeked Hornbill
393	360	Grass Owl	457	424	Grey Hornbill
394	362	Wood Owl	458	425	Redbilled Hornbill
395	361	Marsh Owl	459	426	Yellowbilled Hornbill
396	363	Scops Owl	460	427	Crowned Hornbill
397	364	Whitefaced Owl	461	428	Bradfield's Hornbill
398	365	Pearlspotted Owl	462	429	Monteiro's Hornbill
399	366	Barred Owl	463	430	Ground Hornbill
400	367	Cape Eagle Owl	464	431	Blackcollared Barbet
401	368	Spotted Eagle Owl	465	432	Pied Barbet
402	369	Giant Eagle Owl	466	433	White-eared Barbet
403	370	Pel's Fishing Owl	467	434	Whyte's Barbet
404	371	European Nightjar			(Yellowfronted Barbet)
405	373	Fierynecked Nightjar	468	435	Woodward's Barbet
406	372	Rufoscheeked Nightjar			(Green Barbet)
407	375	Natal Nightjar	469	436	Redfronted Tinker Barbet
408	374	Freckled Nightjar	470	437	Yellowfronted Tinker Barbet
409	376	Mozambique Nightjar	471	438	Goldenrumped Tinker Barbet
410	377	Pennantwinged Nightjar	472	438X	Green Tinker Barbet
411	378	European Swift	473	439	Crested Barbet
412	380	Black Swift	474	440	Greater Honeyguide

18

475	441	Scalythroated Honeyguide	539	515	Whitebreasted Cuckooshrike
476	442	Lesser Honeyguide	540	516	Grey Cuckooshrike
477	442X	Eastern Honeyguide	541	517	Forktailed Drongo
478	443	Sharpbilled Honeyguide	542	518	Squaretailed Drongo
479	444	Slenderbilled Honeyguide	543	519	European Golden Oriole
480	445	Ground Woodpecker	544	520	African Golden Oriole
481	446	Bennett's Woodpecker	545	521	Blackheaded Oriole
482	446X	Specklethroated Woodpecker	546	521X	Greenheaded Oriole
483	447	Goldentailed Woodpecker	547	523	Black Crow
484	448	Knysna Woodpecker	548	522	Pied Crow
485	449	Little Spotted Woodpecker	549	523X	House Crow
486	450	Cardinal Woodpecker	550	524	Whitenecked Raven
487	451	Bearded Woodpecker	551	525	Southern Grey Tit
488	452	Olive Woodpecker	552	—	Ashy Tit
489	453	Redthroated Wryneck	553	526	Northern Grey Tit
490	454	African Broadbill			(Miombo Grey Tit)
491	455	Angola Pitta	554	527	Southern Black Tit
492	456	Melodious Lark	555	528	Carp's Tit
		(Singing Bush Lark)	556	529	Rufousbellied Tit
493	457	Monotonous Lark	557	531	Cape Penduline Tit
494	458	Rufousnaped Lark	558	530	Grey Penduline Tit
495	466	Clapper Lark	559	532	Spotted Creeper
496	468	Flappet Lark	560	533	Arrowmarked Babbler
497	459	Fawncoloured Lark	561	534	Blackfaced Babbler
498	460	Sabota Lark	562	535	Hartlaub's Babbler
499	473	Rudd's Lark			(Whiterumped Babbler)
500	475	Longbilled Lark	563	536	Pied Babbler
501	465	Shortclawed Lark	564	537	Barecheeked Babbler
502	461	Karoo Lark	565	542	Bush Blackcap
503	—	Dune Lark	566	543	Cape Bulbul
504	479	Red Lark	567	544	Redeyed Bulbul
505	464	Dusky Lark	568	545	Blackeyed Bulbul
506	474	Spikeheeled Lark	569	546	Terrestrial Bulbul
507	488	Redcapped Lark	570	547	Yellowstreaked Bulbul
508	490	Pinkbilled Lark	571	548	Slender Bulbul
509	472	Botha's Lark	572	551	Sombre Bulbul
510	491	Sclater's Lark	573	549	Stripecheeked Bulbul
511	492	Stark's Lark	574	550	Yellowbellied Bulbul
512	463	Thickbilled Lark	575	725	Yellowspotted Nicator
513	487	Bimaculated Lark	576	552	Kurrichane Thrush
514	483	Gray's Lark	577	553	Olive Thrush
515	484	Chestnutbacked Finchlark	578	558	Spotted Thrush
516	485	Greybacked Finchlark	579	556	Orange Thrush
517	486	Blackeared Finchlark			(Gurney's Thrush)
518	493	European Swallow	580	557	Groundscraper Thrush
519	494	Angola Swallow	581	559	Cape Rock Thrush
520	495	Whitethroated Swallow	582	560	Sentinel Rock Thrush
521	497	Blue Swallow	583	561	Short-toed Rock Thrush
522	496	Wiretailed Swallow	584	562	Miombo Rock Thrush
523	498	Pearlbreasted Swallow			(Angola Rock Thrush)
524	501	Redbreasted Swallow	585	563	European Wheatear
525	500	Mosque Swallow	586	564	Mountain Chat
526	502	Greater Striped Swallow	587	568	Capped Wheatear
527	503	Lesser Striped Swallow	588	569	Buffstreaked Chat
528	504	South African Cliff Swallow	589	570	Familiar Chat
529	506	Rock Martin	590	571	Tractrac Chat
530	507	House Martin	591	572	Sicklewinged Chat
531	499	Greyrumped Swallow	592	566	Karoo Chat
532	508	Sand Martin	593	573	Mocking Chat
533	509	Brownthroated Martin	594	574	Arnot's Chat
		(African Sand Martin)	595	575	Anteating Chat
534	510	Banded Martin	596	576	Stonechat
535	504X	Mascarene Martin	597	577	Whinchat
536	511	Black Saw-wing Swallow	598	578	Chorister Robin
537	512	Eastern Saw-wing Swallow	599	580	Heuglin's Robin
538	513	Black Cuckooshrike	600	579	Natal Robin

601	581	Cape Robin		662	539	Rockrunner
602	582	Whitethroated Robin				(Damara Rockjumper)
603	593	Collared Palm Thrush		663	617	Moustached Warbler
604	593X	Rufoustailed Palm Thrush		664	629	Fantailed Cisticola
605	591X	Whitebreasted Alethe		665	630	Desert Cisticola
606	589	Starred Robin		666	631	Cloud Cisticola
607	590	Swynnerton's Robin		667	634	Ayres' Cisticola
608	591	Gunning's Robin		668	635	Palecrowned Cisticola
609	592	Thrush Nightingale		669	638	Greybacked Cisticola
610	538	Boulder Chat		670	639	Wailing Cisticola
611	540	Cape Rockjumper		671	641	Tinkling Cisticola
612	–	Orangebreasted Rockjumper		672	642	Rattling Cisticola
613	588	Whitebrowed Robin		673	643	Singing Cisticola
614	583	Karoo Robin		674	644	Redfaced Cisticola
615	586	Kalahari Robin		675	645	Blackbacked Cisticola
616	584	Brown Robin		676	645X	Chirping Cisticola
617	585	Bearded Robin		677	646	Levaillant's Cisticola
618	660	Herero Chat		678	647	Croaking Cisticola
619	595	Garden Warbler		679	648	Lazy Cisticola
620	594	Whitethroat		680	636	Shortwinged Cisticola
621	658	Titbabbler		681	637	Neddicky
622	659	Layard's Titbabbler		682	620	Redwinged Warbler
623	670	Yellowbreasted Hyliota		683	649	Tawnyflanked Prinia
624	668	Mashona Hyliota		684	649X	Roberts's Prinia
625	596	Icterine Warbler		685	650	Blackchested Prinia
626	597	Olivetree Warbler		686	651	Spotted Prinia
627	598	River Warbler				(Karoo Prinia)
628	603	Great Reed Warbler		687	653	Namaqua Prinia
629	–	Basra Reed Warbler		688	619	Rufouseared Warbler
630	604X	European Reed Warbler		689	654	Spotted Flycatcher
631	606	African Marsh Warbler		690	655	Dusky Flycatcher
632	–	Cinnamon Reed Warbler		691	656	Bluegrey Flycatcher
633	607	European Marsh Warbler		692	655X	Collared Flycatcher
634	608	European Sedge Warbler		693	657	Fantailed Flycatcher
635	604	Cape Reed (Lesser Swamp) Warbler				(Leadcoloured Flycatcher)
636	604Y	Greater Swamp Warbler		694	664	Black Flycatcher
		(Rufous Reed Warbler)		695	661	Marico Flycatcher
637	666	Yellow Warbler		696	662	Mousecoloured Flycatcher
638	609	African Sedge Warbler		697	663	Chat Flycatcher
		(Little Rush Warbler)		698	665	Fiscal Flycatcher
639	610	Barratt's Warbler		699	667	Vanga Flycatcher
640	611	Knysna Warbler				(Black-and-White Flycatcher)
641	612	Victorin's Warbler		700	672	Cape Batis
642	616	Broadtailed Warbler		701	673	Chinspot Batis
643	599	Willow Warbler				(Whiteflanked Batis)
644	671	Yellowthroated Warbler		702	675	Mozambique Batis
645	622	Barthroated Apalis		703	674	Pririt Batis
646	623X	Chirinda Apalis		704	676	Woodward's Batis
647	623	Blackheaded Apalis		705	677	Wattle-eyed Flycatcher
648	625	Yellowbreasted Apalis				(Blackthroated Wattle-eye)
649	624	Rudd's Apalis		706	678	Fairy Flycatcher
650	621X	Redfaced Crombec		707	679	Livingstone's Flycatcher
651	621	Longbilled Crombec		708	680	Bluemantled Flycatcher
652	621Y	Redcapped Crombec		709	681	Whitetailed Flycatcher
653	600	Yellowbellied Eremomela		710	682	Paradise Flycatcher
654	626	Karoo Eremomela		711	685	African Pied Wagtail
		(Green Eremomela)		712	688	Longtailed Wagtail
655	602	Greencapped Eremomela		713	686	Cape Wagtail
		(Duskyfaced Eremomela)		714	689	Yellow Wagtail
656	601	Burntnecked Eremomela		715	–	Grey Wagtail
657	627	Bleating Warbler		716	692	Richard's Pipit
658	614	Barred Warbler		717	693	Longbilled Pipit
659	615	Stierling's Barred Warbler				(Nicholson's Pipit)
660	613	Cinnamonbreasted Warbler		718	694	Plainbacked Pipit
661	618	Grassbird		719	695	Buffy Pipit

720	696	Striped Pipit	785	758	Greater Doublecollared Sunbird
721	697	Rock Pipit	786	762	Yellowbellied Sunbird
722	698	Tree Pipit	787	763	Whitebellied Sunbird
723	699	Bushveld Pipit	788	764	Dusky Sunbird
724	700	Short-tailed Pipit	789	765	Grey Sunbird
725	701	Yellowbreasted Pipit	790	766	Olive Sunbird
726	702	Golden Pipit	791	774	Scarletchested Sunbird
727	703	Orangethroated Longclaw	792	772	Black Sunbird
728	704	Yellowthroated Longclaw	793	771	Collared Sunbird
729	–	Fülleborn's Longclaw	794	769	Bluethroated Sunbird
730	705	Pinkthroated Longclaw	795	770	Violetbacked Sunbird
731	706	Lesser Grey Shrike	796	775	Cape White-eye
732	707	Fiscal Shrike	797	777	Yellow White-eye
733	708	Redbacked Shrike	798	779	Redbilled Buffalo Weaver
734	708X	Sousa's Shrike	799	780	Whitebrowed Sparrow-weaver
735	724	Longtailed Shrike	800	783	Sociable Weaver
736	709	Southern Boubou	801	784	House Sparrow
737	–	Tropical Boubou	802	785	Great Sparrow
738	710	Swamp Boubou	803	786	Cape Sparrow
739	711	Crimsonbreasted Shrike	804	787	Greyheaded Sparrow
740	712	Puffback	805	788	Yellowthroated Sparrow
741	731	Brubru	806	789	Scalyfeathered Finch
742	713	Southern Tchagra	807	804	Thickbilled Weaver
743	714	Threestreaked Tchagra	808	790	Forest Weaver
744	715	Blackcrowned Tchagra	809	791X	Oliveheaded Weaver
745	716	Marsh Tchagra	810	791	Spectacled Weaver
		(Blackcap Tchagra)	811	797	Spottedbacked Weaver
746	722	Bokmakierie	812	796	Chestnut Weaver
747	721	Gorgeous Bush Shrike	813	799	Cape Weaver
748	719	Orangebreasted Bush Shrike	814	803	Masked Weaver
749	720	Blackfronted Bush Shrike	815	792	Lesser Masked Weaver
750	717	Olive Bush Shrike	816	801	Golden Weaver
751	723	Greyheaded Bush Shrike	817	800	Yellow Weaver
752	726	Whitetailed Shrike	818	802	Brownthroated Weaver
753	727	White Helmetshrike	819	793	Redheaded Weaver
754	728	Redbilled Helmetshrike	820	854	Cuckoo Finch
755	729	Chestnutfronted Helmetshrike			(Cuckoo Weaver)
756	730	Whitecrowned Shrike	821	805	Redbilled Quelea
757	733	European Starling	822	806	Redheaded Quelea
758	734	Indian Myna	823	807	Cardinal Quelea
759	746	Pied Starling	824	808	Red Bishop
760	735	Wattled Starling	825	809	Firecrowned Bishop
761	736	Plumcoloured Starling			(Blackwinged Bishop)
762	743	Burchell's Starling	826	812	Golden Bishop
763	742	Longtailed Starling	827	810	Yellowrumped Widow
764	737	Glossy Starling			(Cape Bishop)
765	738	Greater Blue-eared Starling	828	816	Redshouldered Widow
766	739	Lesser Blue-eared Starling	829	814	Whitewinged Widow
767	741	Sharptailed Starling	830	815	Yellowbacked Widow
768	740	Blackbellied Starling	831	813	Redcollared Widow
769	745	Redwinged Starling	832	818	Longtailed Widow
770	744	Palewinged Starling	833	829	Goldenbacked Pytilia
771	747	Yellowbilled Oxpecker	834	830	Melba Finch
772	748	Redbilled Oxpecker	835	827	Green Twinspot
773	749	Cape Sugarbird	836	828	Redfaced Crimsonwing
774	750	Gurney's Sugarbird	837	819	Nyasa Seedcracker
775	751	Malachite Sunbird	838	831	Pinkthroated Twinspot
776	752	Bronze Sunbird	839	832	Redthroated Twinspot
777	753	Orangebreasted Sunbird	840	833	Bluebilled Firefinch
778	754	Coppery Sunbird	841	835	Jameson's Firefinch
779	755	Marico Sunbird	842	837	Redbilled Firefinch
780	756	Purplebanded Sunbird	843	836	Brown Firefinch
781	757	Shelley's Sunbird	844	839	Blue Waxbill
782	761	Neergaard's Sunbird	845	840	Violeteared Waxbill
783	760	Lesser Doublecollared Sunbird	846	843	Common Waxbill
784	–	Miombo Doublecollared Sunbird	847	841	Blackcheeked Waxbill

848	842	Grey Waxbill	872	857	Cape Canary
849	842X	Cinderella Waxbill	873	858	Forest Canary
850	825	Swee Waxbill	874	855	Cape Siskin
851	826	East African Swee	875	856	Drakensberg Siskin
852	844	Quail Finch	876	861	Blackheaded Canary
853	845	Locust Finch	877	863	Bully Canary
854	838	Orangebreasted Waxbill	878	866	Yellow Canary
855	821	Cut-throat Finch	879	865	Whitethroated Canary
856	820	Redheaded Finch	880	869	Protea Canary
857	823	Bronze Mannikin	881	867	Streakyheaded Canary
858	824	Redbacked Mannikin	882	868	Blackeared Canary
859	822	Pied Mannikin	883	875	Cabanis's Bunting
860	846	Pintailed Whydah	884	874	Goldenbreasted Bunting
861	847	Shaft-tailed Whydah	885	873	Cape Bunting
862	852	Paradise Whydah	886	872	Rock Bunting
863	853	Broadtailed Paradise Whydah	887	871	Larklike Bunting
864	849	Black Widowfinch	901		Mountain Pipit
865	850	Purple Widowfinch	902		Lesser Yellowlegs
866	–	Violet Widowfinch	903		Redthroated Pipit
867	851	Steelblue Widowfinch	904		Redrumped Swallow
868	870	Chaffinch	905		Laysan Albatross
869	859	Yelloweyed Canary	906		Greater Yellowlegs
870	860	Blackthroated Canary	907		Pied Wheatear
871	860X	Lemonbreasted Canary			

Reproduced with the kind permission of the Southern African Ornithological Society

FAMILY INTRODUCTIONS

The following are general accounts of the major bird families represented in southern Africa.

Penguins. Family **Spheniscidae**
Flightless marine birds. Wings are powerful, rigid flippers used for rapid underwater swimming. Although, when swimming, penguins are difficult to see as they hold their bodies very low in the water, they often 'porpoise', seeming like large fish. Ashore they waddle with a clumsy side-to-side gait or hop. The stiff, dense feathers are shed in a complete annual moult during which the bird comes ashore for a period of three to four weeks. Most vagrants to our area are birds which are either ill or have been forced ashore to moult. Eight species occur in this region: four in the southern ocean, and of the four in southern Africa, the Jackass Penguin is endemic.

Grebes. Family **Podicipedidae**
Small to medium-sized diving birds with lobed toes (not webbed feet), a straight, pointed bill in the larger species, and head feathering which in the breeding season becomes richer in colour and more elaborate. Feed on aquatic animals caught underwater, with the larger species feeding almost entirely on fish. Nests are mounds of rotting vegetation placed on floating matter or amongst reeds. Young have striking head patterns and are often seen riding on their swimming parent's back. Local migrations are undertaken. Three species occur in our region.

Albatrosses. Family **Diomedeidae**
Huge, long-winged seabirds which come ashore only to breed, sometimes in large colonies on small oceanic islands, chiefly in the southern hemisphere. Identifications are based on underwing patterns and bill colours. Closely related to the shearwaters and tiny storm petrels, they share the peculiar feature of separate, raised tubular nostrils. Prodigious fliers, they glide on their long, narrow, stiffly-held wings in high winds. However, in calmer conditions they have a flapping flight and also spend long periods resting on the water. Large numbers are encountered over the deep-sea trawling grounds in the south-west. Nine species occur within our region.

Shearwaters and petrels. Family **Procellariidae**
The genera in this family vary in size from the small prions to the huge giant petrels. The *Puffinus* group differs from the *Pterodroma* group by having long, slender (not short and stubby) bills and straight, stiffly-held wings, not angled and flexed at the wrist. All the species are tube-nosed. Breed on remote oceanic islands in both hemispheres. Individual species have characteristic flight actions which aid identification. Sexes are usually alike and most species do not show an obvious immature plumage. Although many are attracted to ships to feed on galley scraps and offal, others, especially the *Pterodroma* petrels, avoid or are indifferent to them. Twenty-five species occur in our region.

Storm petrels. Family **Oceanitidae**
The smallest members of the tube-nosed group of pelagic seabirds. Very swallow- or swift-like in appearance, they tend to intersperse their gliding with hovering and fluttering over the water, and often dabble their feet. In continuous flight, they stay very close to the surface of the water. During prolonged bad weather and in times of short food supply, they may venture very close inshore. Small, loose colonies breed on remote oceanic islands in both hemispheres. Seven species occur in our region, of which five have been seen in southern African waters.

Diving petrels. Family **Pelecanoididae**
Small, dumpy, short-winged seabirds confined to the southern hemisphere. Flight is extremely rapid on whirring wings, with the birds appearing to fly through wave tops. When feeding, they fly swiftly, stop abruptly, then flop on to the surface, disappearing instantly as they dive in search of small crustaceans. Impossible to identify in the field because of their similarity. Two species occur in the region.

Tropicbirds. Family **Phaethontidae**
Mainly white, medium-sized seabirds identifiable by combination of bill and tail streamer colour, and by the amount of black shown in their wings. Sexes are alike. Immatures are heavily streaked and

barred, and they lack the long tail streamers of adults. Tropicbirds plunge-dive to obtain their food, mainly flying fish. They breed under bushes or in clefts of cliffs on tropical islands. Two species are vagrant to our region.

Pelicans. Family **Pelecanidae**
Large, heavily built birds with grey, black and white plumage. The exceptionally long bill has a distensible pouch which is used as a scoop net when catching fish and is not, as popularly believed, used to store food. Identifiable by the contrast of black and white on their wings and the bill base structure. Found in both salt- and freshwater localities. Although ponderous in flight, pelicans are dynamic gliders. Very ungainly on land. They breed colonially either in trees or on the ground. Two species occur in our region.

Gannets and boobies. Family **Sulidae**
Large white, or white and brown seabirds with cigar-shaped bodies and hefty, pointed bills. Inshore or open ocean feeders, they plummet from considerable heights into the ocean to pursue fish. Colonial breeders on islands and cliffs. In our region, three species occur: one is an endemic breeder.

Cormorants. Family **Phalacrocoracidae**
Small to large, black, or black and white birds with long necks and tails. Cormorants forage inland and in marine coastal waters, diving from the surface to catch fish with their long, hooked bills. Commonly seen with wings outstretched, which helps to dry the wings and keep the bird warm. Colonial breeders on rocky islands, cliffs and in trees. Six species occur in our region, three of which are endemic.

Darters. Family **Anhingidae**
Medium-sized birds with very thin, long necks and elongated tails. Using their dagger-shaped bills, they spear fish under water, much like herons do from the surface. When swimming, the body is held submerged and, with only the slender neck and head visible, a darter resembles a swimming snake, hence the name 'snakebird'. Colonial breeders in trees. The young are covered in white, fluffy down. One species occurs in our region.

Frigatebirds. Family **Fregatidae**
Large, very long-winged and fork-tailed seabirds. Adapted to an aerial way of life, they rarely perch and never swim or walk. In the course of their superb soaring flight high over the ocean, they descend swiftly to harass boobies and terns, forcing them to disgorge their last meal, which the frigatebirds then retrieve. Colonial breeders on remote tropical islands. One species is a vagrant to our region.

Herons and bitterns. Family **Ardeidae**
Herons are medium- to large-sized waders with long legs and elongated, slender necks. Plumage is variable but many species are white and are usually referred to as egrets. Bitterns are generally more squat, more furtive, hiding in reedbeds, and have unusual, loud booming calls. In flight, the head is tucked well into the body. Herons nest colonially in reedbeds or trees, whereas bitterns nest solitarily in reedbeds. Nineteen species occur in our region.

Hamerkop. Family **Scopidae**
A strange-looking, medium-sized brown bird with a laterally compressed bill and shaggy, elongated nape feathers. It is neither a stork nor a heron, but shows characters of both. Builds an enormous domed nest of mud, sticks and grass in a tree. The Hamerkop is the only member of its family and is confined to Africa and Madagascar.

Storks. Family **Ciconiidae**
Large birds with long necks and legs, relatively short tails, and very broad wings. Most species have striking black and white plumage. The Openbilled Stork has a gap between its mandibles which enables it to manoeuvre the bivalves on which it feeds. Storks nest singly or semi-colonially in trees or on cliffs. Eight species occur in our region.

Ibises and spoonbills. Family **Plataleidae**
Medium-sized birds with elongated, decurved or flattened bills, long legs and variable plumage coloration. They feed by probing in shallow water, mud or grass. Spoonbills feed by moving their bills

from side-to-side in shallow water, thus skimming off aquatic animals. Five species occur in our region, one of which is endemic.

Swans, ducks and geese. Family Anatidae
Small to large freshwater-dwelling birds with blunt, flattened bills. They either dive or up-end in the water in search of food. Geese regularly graze grass in open fields. Flight is fast and direct with the neck held outstretched. Nests, built on the ground or in holes in trees, are lined with down plucked from the female's breast. The introduced Mute Swan is almost extinct in the wild, although records of escaped exotic wildfowl are regular. Twenty species occur in the region, two of which are endemic.

Vultures, eagles, buzzards, goshawks and sparrowhawks, and harriers. Family Accipitridae
Small to huge birds of prey. Many species have streaked or barred underparts and plumage is usually grey or brown. Most are well adapted for catching prey, with hooked bills and powerful feet which end in sharp toenails. They scavenge or hunt a wide variety of prey, from insects to small antelope.

Vultures.
Large birds with long, very broad wings designed for soaring. The heads of most are unfeathered to varying degrees. These scavengers gather around carcasses, fighting and jostling as they rip apart the hide with powerful, hooked bills. Eight species occur in the region, one of which is endemic.

Eagles.
Medium to large birds of prey with long, broad wings and feathered legs, which distinguish them from Snake Eagles and Buzzards. They are noted for their soaring flight and their hunting prowess. Thirteen species occur in the region.

Buzzards.
Although similar to eagles in shape, buzzards are generally smaller and have unfeathered legs. Typical call is a cat-like mewing. Plumages are very variable, making identification complex. Flight is laboured with much soaring but little hovering. Buzzards prey on insects, lizards, and small birds and mammals. Mostly tree nesters, they frequent well-wooded areas and open plains. Six species occur, one of which is endemic.

Goshawks and sparrowhawks.
Small to medium-sized birds with rounded wings and long tails. Most species have button-like yellow or red eyes and some have very long toes designed for gripping their prey of birds and small mammals. They hunt by a dash-and-seize technique. Twelve species occur in the region, one of which is endemic.

Harriers.
Medium-sized birds with long, narrow wings and tails. They are distinctive in flight as they glide low over the ground, head down, with wings held in a shallow 'V'. In such flight, they suddenly stall, flopping into the grass to grasp their prey, which includes frogs and small rodents. Mostly seen over marshes and open fields. Five species occur, one of which is endemic.

Falcons and kestrels. Family Falconidae
Small to medium-sized birds, they are swift, agile fliers on their very long, pointed wings. Some species are dynamic aerial hunters, stooping at great speeds to strike at their prey in mid-air. The female is often much larger than the male. Sixteen species occur in the region.

Francolins, quail, buttonquail and guineafowl. Families Phasianidae, Turnicidae and Numididae
Rotund, small to medium-sized gamebirds, identified mainly by their head and breast markings. Except for quails, they are gregarious birds, and all have distinctive crowing calls. All species are terrestrial although some do roost in trees. Large clutches are laid in nests on the ground, well concealed by vegetation. Flight, usually over short distances, is fast on whirring wings. Twenty species occur in the region, five of which are endemic.

Rails, crakes, gallinules, moorhens and coots. Family Rallidae
Small to medium-sized, ground-dwelling birds with long legs and toes. The very short tail is often held erect and flicked up and down. Good swimmers, they generally inhabit marshes and, notwithstanding their furtive behaviour, can frequently be seen clambering through reeds. Although most are reluctant to take flight, many species undertake long-distance migrations at night. In our region 21 species occur, one of which is a vagrant from the Americas, and a further two species are endemic to Gough and the Tristan group of islands.

Cranes. Family Gruidae
Very large, long-legged birds with relatively short bills. They inhabit wetlands and open grasslands and, when not breeding, congregate in large flocks. All have highly complex dancing displays and are extremely vocal. They have a slow, laboured flight during which the long neck is held outstretched. Three species occur, one of which is endemic and is South Africa's national bird.

Bustards. Family Otididae
Medium to very large terrestrial birds, with long, sometimes very slender necks, and long legs. Variable in colour but usually cryptically mottled buff, brown and black above. Walking is slow, with the neck being swung back and forth. Reluctant to take flight, the birds tend to crouch or run when alarmed. The male performs an elaborate courtship ritual in which he fluffs out his neck feathers and executes aerial displays while giving raucous calls or whistles. Ten species occur, six of which are endemic.

Jacanas. Family Jacanidae
Small, freshwater-dwelling birds with elongated toes and nails which distribute their weight and allow them to walk on floating vegetation. They also nest on floating vegetation. The males, which are smaller than the females, take the major role in the incubation and rearing of the young. Seldom seen swimming, they fly strongly from one area of floating weeds to another. Two species occur in the region.

Waders: plovers, sandpipers, phalaropes, curlew and godwits. Families Charadriidae, Scolopacidae, Phalaropodidae, Haematopodidae, Rostratulidae, Recurvirostridae and Dromadidae
A large group of wading birds, varying greatly in size from very small to medium, and from the long-legged to the very short-legged. The larger species are normally easy to identify but the smaller species are more difficult and are usually distinguished by a combination of their wing and rump patterns, and by bill shape and length. The majority of waders occurring in the region breed in the northern hemisphere, usually the Arctic regions, and spend the summer months on our estuaries and in freshwater regions. Most have bills adapted for foraging in muddy areas where they probe and pick for small invertebrates. Fifty-eight species have been recorded in the region, 42 of which are visitors. Only one is endemic.

Dikkops. Family Burhinidae
Medium-sized, cryptically coloured wader-like birds with large heads. Their large, yellow eyes are indicative of their nocturnal habits. Generally quiet during the day, at night their mournful cries and whistles can be heard over great distances. They occur in the vicinity of water courses in dry woodlands to semi-desert. Two eggs are laid in a scrape on the ground. Two species occur in the region.

Coursers and pratincoles. Family Glareolidae
Coursers are wader-like birds which generally inhabit the drier regions. Their long legs enable them to run swiftly and their cryptic back coloration blends well with their sandy environment. In flight they show boldly patterned wings. Pratincoles are short-legged, plover-like birds with long, pointed wings and black and white forked tails. When in flight, the combination of these characteristics results in their resembling huge swallows. Both groups have large eyes, indicative of their nocturnal and crepuscular habits. Insectivorous, they catch their prey in flight or on the ground. Eight species occur, one of which is endemic.

Skuas. Family Stercorariidae
Medium to large, brown or brown and white gull-like seabirds. The smaller species are often difficult to

identify and, in this regard, combinations of size and shape are important. Predatory, they chase gulls and terns, forcing them to regurgitate food. Fish, crustaceans, young birds and eggs also form part of their diet. When harrying other seabirds they display amazing speed and agility, and the smaller species often resemble falcons when in fast pursuit. They breed mostly in the higher latitudes of both Poles. Five species occur in the region.

Gulls and terns. Family Laridae
Small to large waterbirds which frequent inland and coastal regions. Predominantly grey and white birds, with distinct immature plumages, gulls are usually identifiable by their wing and head patterns, and bill and leg coloration. Fork-tailed, terns are on average smaller than gulls and have a more bouyant, agile flight on long, pointed wings. When breeding, terns display brightly coloured bills and black caps. Thirty-one species occur in the region, one of which is an endemic breeder.

Sandgrouse. Family Pteroclididae
Desert and dry woodland-dwelling birds with cryptically coloured plumage which makes them difficult to see when at rest in their arid environment. Every day they fly considerable distances to reach their water sources (each species preferring different drinking times), sometimes gathering in their thousands to drink. On the ground they look like very short-legged francolins but in flight they resemble swiftly flying doves. Four species occur, two of which are endemic to the region.

Pigeons and doves. Family Columbidae
The term 'pigeon' normally refers to the larger species and 'dove' to the smaller members of this family. Most have distinctive calls and are identifiable by their tail patterns and various other plumage colorations. All domestic pigeons are descendants of the Rock Dove of the northern hemisphere. Grain seeds comprise the bulk of their diet but most forest doves are fruit-eaters. Fourteen species occur.

Parrots and lovebirds. Family Psittacidae
Small to medium-sized birds with vivid green, brown and red coloration. The beak is short, stubby and deeply hooked, ideally adapted for cracking hard nuts and ripping open fruit. Flight is rapid and direct. Calls consist of shrieks and screams; in the wild these species do not mimic the human voice as they do in captivity. Eight species occur, of which two are endemic. Another species, an aviary escapee, has set up a viable feral population.

Louries. Family Musophagidae
Medium-sized, long-tailed birds which, except for the drab Grey Lourie, display bright green, red and blue plumage. Mostly forest dwellers, they have loud, raucous calls and show a distinctive bounding action as they leap from branch to branch through the canopy. Flight is laboured, with fast wing beats interspersed with long glides. Their diet consists mainly of fruit. Four species occur in the region.

Cuckoos and coucals. Family Cuculidae
Small to medium-sized birds that range in colouring from drab greys to gaudy greens and yellows. With their long, pointed wings, elongated tails and barred underparts, the larger species appear very falcon-like in flight. Cuckoos are brood parasites, but coucals rear their own young. All have distinctive calls. Eighteen species occur in the region.

Owls. Families Tytonidae and Strigidae
Small to large nocturnal birds of prey. All have distinctive calls in the form of hoots and shrieks, and are most vocal just after dusk and before dawn. The Barn and Grass owls (Tytonidae) differ from other owls (Strigidae) by their heart-shaped facial discs of pale, stiff feathers surrounding their small dark eyes. The plumage of both is soft and fluffy with brown, buff or grey colouring, often with heavy barring or streaking. Owls are very silent fliers and their prey ranges from insects to mammals and birds and, in one species, fish. In general, owls are more easily located at night, initially by their calls, and then observed by torchlight. Twelve species occur.

Nightjars. Family Caprimulgidae
Nocturnal, dove-sized birds with large heads and eyes, and a wide gape for catching insects in flight. Cryptically coloured, they are difficult to locate during the day because they are superbly

camouflaged as they rest in leaf litter and stony areas. Although it is not easy to distinguish between the species, all have distinctive amounts of white in wings and tail. As with owls, nightjars are best identified when heard calling at night, and then tracked by torchlight. Seven species occur.

Swifts. Family Apodidae
Their long, sickle-shaped wings make the flight of swifts rapid and effortless. They do not perch, but feed on insects caught while on the wing, and are even thought to roost in flight at great heights. Their legs are very short and the toes all point forward, an adaptation for clinging to their nesting places: rock faces, tree bark or palm fronds. Identification is based largely on size, rump pattern and tail shape. Thirteen species occur, one of which is a vagrant and another, an endemic.

Kingfishers. Family Halcyonidae
Very small to medium-sized birds, frequently dazzling blue and red in colour, and with long, stout, pointed and often brightly coloured bills. They frequent water and woodland habitats, feeding on fish, insects and reptiles. Fish-eating species often hover before plunge-diving to capture their prey, whereas insect- and reptile-eaters sit motionless on a perch before dashing after their prey with a fast, direct flight. Ten species occur.

Bee-eaters. Family Meropidae
A group of brightly coloured, slender birds that usually occur in flocks and have distinctive contact calls. They have long, slightly decurved bills, long pointed wings and some species also have elongated central tail feathers. Insectivorous, they hawk their prey from the ground or exposed perches and are attracted to veld fires, where they glean insects flushed by the heat. Some are colonial breeders, nesting at the end of long tunnels excavated in sandy banks. Most are identified by their brilliant colour combinations and tail projections. Eight species occur.

Rollers. Family Coraciidae
Stocky perching birds which derive their common name from their acrobatic, noisy display flights during which they tumble and roll through the air. Most species have a plumage combination of bright blues, greens, violets and browns. All nest in holes in trees. Although mostly insectivorous, they also eat reptiles and small rodents. Five species occur.

Woodhoopoes and hoopoe. Families Phoeniculidae and Upupidae
The woodhoopoes are glossy black, purple and green birds with long, decurved bills, elongated, pointed tails, and varying amounts of white in the wings and tail. Found in noisy family groups, they are co-operative breeders, with the whole group attending to the raising of the young. The Hoopoe differs from the woodhoopoes by lacking the metallic sheen to the feathers, and by having a long crest and a boldly patterned black and white back and wings. Four species occur, with one endemic to our region.

Hornbills. Family Bucerotidae
Medium-sized to large birds with long, heavy, decurved bills; some species have a casque on the upper mandible. Most are identifiable by bill and body coloration. While incubating the eggs, the female is sealed inside the nest cavity by the male, and during this period she moults her flight feathers. Most hornbills are woodland inhabitants and their diet consists mainly of fruit and berries. However, an exception to this is the huge Ground Hornbill, which feeds on insects, reptiles and small rodents. Nine species occur in the region, two of which are endemic.

Barbets. Family Capitonidae
Small, woodpecker-related birds with stout bodies and large heads and bills. They have an unusual toe arrangement in that two toes point forward and two back. All have distinctive calls, the tinker barbets having clinking call notes. Frugivorous and insectivorous forest- and bush-dwellers, they use their stout bills to excavate nesting holes in dead trees. Ten species occur, one of which is endemic.

Honeyguides. Family Indicatoridae
Small, short-legged birds, honeyguides are usually drab in coloration and have short, stubby bills. Species identification is usually based on call, shape of bill, and habits. Larger species are known to lead mammals, especially the honey badger and man, to beehives in the hope of sharing the spoils of a raided nest. They are unique in that they eat beeswax. Brood parasites, they lay their eggs in the nests of barbets and woodpeckers. Six species occur.

Woodpeckers and wryneck. Families **Picidae** and **Jyngidae**
Small to medium-sized arboreal birds with stout, pointed bills which are used to hammer and bore into wood to reach grubs and insects and to excavate nest holes. The tails are stiff and pointed and brace the birds as they cling to branches and move jerkily up tree trunks. Most are identifiable by breast and back markings and by the amount of red on the head. One species, the endemic Ground Woodpecker, has adapted to a terrestrial habitat. Nine other species occur, including another endemic.

Larks. Family **Alaudidae**
Small and nondescript terrestrial birds which resemble pipits but have shorter tails, stouter bills and dumpier bodies. Most species frequent open areas in the drier regions and are cryptically coloured to blend with their stony and sandy habitats. Those species that inhabit woodland and bush areas are normally heavily streaked above and below. Identification is based on subtleties of plumage coloration, as well as bill shape, and song. All build their nests on the ground, some being well concealed, others a mere scrape. Twenty-six species occur, 18 of which are endemic.

Swallows and martins. Family **Hirundinidae**
Shorter- and less stiff-winged than swifts, they are frequently seen perched on wires or reeds. Plumage colours range from glossy blues to reds and dull browns. All are aerial and insectivorous and for this reason have flattened bills with a very wide gape. Some are colonial breeders, nesting in holes in riverbanks, or using mud pellets to build elaborate nests on man-made structures. Twenty-one species occur, one of which the South African Cliff Swallow, is a breeding endemic.

Cuckooshrikes. Family **Campephagidae**
Slow-moving, arboreal birds which superficially resemble the cuckoos. Insectivorous, they glean the canopy in the evergreen forests they frequent and have soft calls. Sexual plumage differences are very slight except in the Black Cuckooshrike where the female is heavily barred green, yellow and brown. Three species occur.

Drongos. Family **Dicruridae**
Noisy, black insectivores. Bold and conspicuous, they are known to be fearless when harassing large birds of prey, giving chase and dive-bombing them in flight. They have stout, slightly hooked bills with prominent rictal bristles and they catch insects on the wing, darting from an exposed perch in various wooded and open country habitats. Two species occur.

Orioles. Family **Oriolidae**
Resembling starlings in size and shape, these highly coloured birds of forests and bushveld all have very similar clear and fluty calls. The males are a vivid yellow and are not easily located in their leafy surroundings. Females have drabber green and yellow plumages. They feed on a variety of fruits and insects and occasionally, nectar. Four species occur, one of which, the Greenheaded Oriole, is confined to a mountain top in Moçambique.

Crows and ravens. Family **Corvidae**
Fairly large, black or black and white birds with strong bills. At close range, it can be seen that the black parts of the plumage are in fact glossy purple and green. The calls of these notorious scavengers are throaty grunts and raucous cries but, in captivity, they can be trained to mimic the human voice. They are found in a variety of habitats throughout the region and the Pied Crow has adapted successfully to cities. Four species occur, with the House Crow, a recent arrival, having established itself in Natal.

Tits. Family **Paridae**
Small, highly active arboreal birds with short, robust bills, and black, white and grey plumages. Usually seen in small groups of four to six, they are prominent and noisy members of mixed bird parties foraging through tree canopies. Mostly insectivorous, they nest in holes in trees or in old drainpipes. Six species occur, three of which are endemic.

Babblers. Family **Timaliidae**
Very vocal and sociable thrush-like birds which congregate in small groups and continually chatter to each other to keep contact. Insectivorous, they feed mostly on the ground but seek refuge in trees or reedbeds. All are very distinct and easily identifiable. Five species occur, two of which are endemic.

Bulbuls. Family Pycnonotidae
Possibly one of the most familiar groups of birds in the region. They are often abundant in suburbia, although the majority are extremely shy and skulking forest- and bush-dwellers and are consequently very difficult to see. Their diet consists of fruit and insects and one species specializes in gleaning the leaf litter for food. They are usually the first birds to make a noise when a snake or an owl is found concealed in a tree. Eleven species occur, three of which are endemic.

Thrushes and their allies. Family Turdidae
This family comprises the thrushes, robins and chats, all of which are very varied in plumage coloration and habitat preference. True thrushes, the larger species, are forest-floor dwellers, retiring to thicker tangles when disturbed. Most robins are shy, skulking birds and have clear melodious songs, which often mimic those of other birds. Chats frequent more open country, are often longer-legged than robins, have a very upright stance, and continually flick their wings. Forty-three species occur, 19 of which are endemic.

Warblers. Family Sylviidae
A very varied group of small, active, insectivorous birds whose habitats range from forests to reedbeds. Identification in the field is often highly complex; however, song is a very reliable character and in some species, especially the cisticolas, provides the only sure means of distinction. The sexes are alike in most species. Seventy-two species occur, 17 of which are endemic.

Flycatchers. Family Muscicapidae
Small birds with broad, flattened bills. They hawk insects in flight by sallying forth from an exposed perch to which they often return to repeat the performance. Some species forgo this flycatching technique and, like warblers, glean insects from foliage. Plumage and behaviour are highly variable: most species are sombre coloured and quiet, while others are vivid and have distinctive calls and songs. Twenty-one species occur, six of which are endemic.

Wagtails, pipits and longclaws. Family Motacillidae
Small birds, most wagtails and pipits have very long tails which they constantly 'wag'. The pipits are confusingly similar, streaked and blotched brown birds which are very difficult to identify until their calls and songs are learnt. Longclaws are larger and show distinctive throat and breast markings. All frequent grassland areas, the wagtails normally in association with water. Twenty-one species occur, three of which are endemic.

Shrikes, bushshrikes and helmetshrikes. Families Laniidae, Malaconotidae and Prionopidae
Small to medium-sized birds, very variable in colour but most are black, white and grey. The true shrikes are long-tailed birds with heavy, hooked bills. They are conspicuous, aggressive birds that hunt their prey (mostly insects) from a prominent perch, and frequently impale their victims on thorns or barbed wire. Although bushshrikes are shy, furtive birds, they are often brightly coloured and have very distinct, ringing calls. The helmetshrikes are distinctive with their stiff, bristly forehead feathers. Twenty-six species occur, seven of which are endemic.

Starlings. Family Sturnidae
Small to medium-sized birds, some have long tails, and most display glossy blue and green plumage. Insectivorous and frugivorous, most are both terrestrial and arboreal in their habits. Some species are very gregarious and form large flocks, especially to roost. Most are very distinct, presenting few identification problems. Fourteen species occur, three of which are endemic.

Oxpeckers. Family Buphagidae
Starling-related birds with bright yellow and red bills, and feet with a powerful grip, adapted for clinging to mammals' fur. Small groups gather and feed on ticks found on large game and cattle. Seldom seen outside game reserves since tank-dipping has rid most livestock of ticks. Two species occur: although locally common in the north, the Yellowbilled Oxpecker has recently become extinct in South Africa.

Sugarbirds. Family Promeropidae

Their noisy, chattering calls indicate that they are related to starlings. Bills are long and decurved, well designed for probing protea and aloe blooms in search of nectar and for obtaining insects attracted to the flowers. They have elongated tails, yellow vents and varying amounts of russet on the head. Both species that occur are endemic: this is the region's only endemic family.

Sunbirds. Family Nectariniidae

Most of the males of these small birds have brilliant plumage and are easily identifiable but the females are drabber and can be confusingly similar. The long, decurved bills are ideal for probing flowers for nectar but the birds also eat insects to supplement their diets. When probing flowers, they usually perch, rarely hovering. Their flight is fast and dashing. Twenty-one species occur, four of which are endemic.

White-eyes. Family Zosteropidae

Small, arboreal birds which have short, stout bills, yellow and green plumage, and white rings around dark brown eyes. Gregarious, they behave like tiny babblers, gathering in small parties and continually calling to keep contact. They are both insectivorous and frugivorous, occasionally also feeding on nectar. Two species occur, one of which is endemic.

Weavers, sparrows and allies. Family Ploceidae

Small birds with thick, robust bills designed for their diet of grain and grass-seeds. In their highly coloured breeding plumages, the male weavers, widowbirds and bishops are easily identifiable, but during non-breeding periods they resemble the females and are almost indistinguishable except to the trained observer. Most weavers build intricate nests and the structure and weaving are distinctive. When not breeding, they gather in large flocks to forage over a wide area, congregating to roost, usually in reedbeds. Thirty-four species occur, four of which are endemic.

Waxbills and allies. Family Estrildidae

Small, very brightly coloured birds, often with red or blue bills that have a waxy sheen. Most are seed-eaters but feed insects to their young. All are terrestrial feeders but many are shy, skulking birds, and when alarmed, will dive for cover in thick bushes. Many play host to the parasitic whydahs and widow-finches. Twenty-seven species occur, three of which are endemic.

Whydahs and widowfinches. Family Viduidae

Small, pied or all-black seedeaters. Although the males are distinctive in breeding plumage, in non-breeding plumage they resemble the females and immatures which, in most species, are almost indistinguishable. Brood parasites on the waxbill family in the breeding season, they form flocks when not breeding. Eight species occur, one of which is endemic.

Finches, canaries and buntings. Family Fringillidae

Small birds, they range from the brilliantly coloured to the drab and dowdy. Most are seed- and insect-eaters and are terrestrial in habits, but use trees and scrub for nesting. Highly prized by the cagebird industry for their fine, buzzy songs. Twenty-three species occur, twelve of which are endemic.

2 **King Penguin** *Aptenodytes patagonicus* L 94 cm
The long pointed bill, large size and bright yellow ear patches are distinctive. Imm. is a paler version of ad. *Habitat.* Breeds on sub-Antarctic islands. Pelagic and ranges throughout southern oceans south of 40° S. Extreme vagrant: one record from the southern Cape, possibly a ship-assisted bird. *Call.* Short bark or loud trumpeting during display. *Afrikaans.* Koningspikkewyn.

3 **Jackass Penguin (Cape Penguin)** *Spheniscus demersus* L 60 cm
The only resident penguin in our area, it has a diagnostic black and white facial pattern. Imm. is dark greyish blue with grey cheeks and an incomplete grey chest band. Some birds show a double bar on the throat and chest – this is diagnostic of the Magellanic Penguin *S. magellanicus* of Argentina which has not been positively recorded in our region. *Habitat.* Occurs within 50 km of the shore and breeds on islands off the Cape and Namibian coasts. Endemic, and common but declining. *Call.* A loud, donkey-like braying, especially at night. *Afrikaans.* Brilpikkewyn.

4 **Rockhopper Penguin** *Eudyptes chrysocome* L 61 cm
Ad. has a short, stubby red bill and a pale yellow stripe extending from in front of the eye to the nape, where it ends in a shaggy crest. The eyebrow stripe does not meet on the forehead as in the Macaroni Penguin.
Habitat. Breeds on sub-Antarctic islands, and is pelagic south of 30° S when not breeding. Vagrant. *Call.* Display call is a trumpeting 'wada, wada, wada'. *Afrikaans.* Geelkuifpikkewyn.

5 **Macaroni Penguin** *Eudyptes chrysolophus* L 70 cm
Distinguished from Rockhopper Penguin by stouter, more robust bill, larger body size, and orange-yellow eyebrows that meet on the forehead. At sea differentiated from Rockhopper by pale pink patch on sides of gape and white spot on rump. Imm. differs from Rockhopper by yellow eyebrow starting above the eye and not the more pronounced yellow stripe that starts before the eye in Rockhopper. *Habitat.* At sea and on sub-Antarctic islands. Rare vagrant with three records from our region. *Call.* A loud trumpeting when in display. At sea a harsh 'aaark'. *Afrikaans.* Macaronipikkewyn.

6 **Great Crested Grebe** *Podiceps cristatus* L 50 cm
Breeding ad. unmistakable, having a dark double crest with rufous fringes on sides of head. In winter shows a dark cap contrasting with silvery cheeks, throat and neck. Body held low in water with neck very erect. In flight shows a pale wing bar. *Habitat.* Large open stretches of fresh water. *Call.* Mostly silent but utters soft calls during display. *Afrikaans.* Kuifkopdobbertjie.

7 **Blacknecked Grebe** *Podiceps nigricollis* L 28 cm
Far smaller than Great Crested Grebe but larger than Dabchick. Breeding ad. has black head and throat, and conspicuous chestnut ear tufts. Winter ad. distinguished from Dabchick by larger size, white cheeks and throat. At close range the cherry red eye is noticeable. *Habitat.* Open stretches of fresh water, especially coastal sewage farms. When non-breeding, common on inshore waters of the Cape and Namibia. Breeds widely on well-vegetated dams. *Call.* Mostly silent but known to utter a trill. *Afrikaans.* Swartnekdobbertjie.

8 **Dabchick (Little Grebe)** *Tachybaptus ruficollis* L 20 cm
When breeding, this very small, dark grebe has chestnut sides to the neck, and a diagnostic pale creamy spot at the base of the bill. In winter, distinguished from Blacknecked Grebe by smaller size, dusky cheeks and throat. *Habitat.* Virtually any open stretch of fresh, still water. *Call.* A very readily uttered whinnying trill. *Afrikaans.* Kleindobbertjie.

2▲   3▲   4▲

5▲   7▼   8▼   6▲

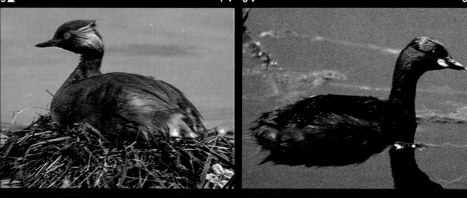

9 Royal Albatross *Diomedea epomophora* L 122 cm
Only distinguishable at close range from Wandering Albatross by black eyelids and black cutting edge to pinkish mandibles. The race *sandfordi* is differentiated by having a thick black border to the underwing, from carpal joint to wing tip. Imm. resembles ad. and lacks the brown plumage of imm. Wandering Albatross. *Habitat.* Southern oceans. Rare vagrant to southern Cape seas. *Call.* Not recorded in our region. *Afrikaans.* Koningmalmok.

10 Wandering Albatross *Diomedea exulans* L 125 cm
The white back distinguishes the ad. from all except the Royal Albatross. Imm. in first plumage is dark with white face and underwing. With age plumage pales and progresses through various spotted and mottled stages. *Habitat.* Southern oceans south of 30° S. Breeds on sub-Antarctic islands. Visits our coastal waters chiefly in winter, ranging widely off the Cape and Natal coasts. Avoids trawling grounds of the western Cape. *Call.* Harsh, nasal 'waaaak'. *Afrikaans.* Grootmalmok.

11 Shy Albatross *Diomedea cauta* L 98 cm
Largest of the 'black-backed albatrosses'. The upperparts are paler than in other albatrosses and the underwing is white with a very narrow black border. Ad. bill is olive with a yellow tip, in imm. it is greyish with a black tip. Imm. Blackbrowed Albatross also has a grey bill with a black tip but is a smaller bird and has a dark, not white underwing pattern. Ad. Shy Albatross has greyish head with white crown; imm. has varying amounts of grey on head and has smudges on sides of chest. *Habitat.* Southern oceans. Common visitor to our coasts, chiefly in winter. *Call.* 'Waak', a loud, raucous call. *Afrikaans.* Bloubekmalmok.

12 Blackbrowed Albatross *Diomedea melanophris* L 90 cm
Ad. has a yellow bill with a pinkish tip. Imm. has grey bill with black tip, and varying amounts of grey on head and sides of neck which sometimes join to form a collar. Underwing in first year plumage is all dark but pales with age. Underwing of imm. Greyheaded Albatross is also dark but Greyheaded has an all-dark bill. Imm. Shy and Yellownosed albatrosses have mostly white underwings with black borders. *Habitat.* Southern oceans. Visitor to all our coasts throughout the year but more common in winter. The most common albatross in Cape waters. *Call.* Grunts and squawks when squabbling over food. *Afrikaans.* Swartrugmalmok.

13 Greyheaded Albatross *Diomedea chrysostoma* L 89 cm
Similar in size to Blackbrowed Albatross but differs by having a grey head and a black bill with yellow stripes along the upper and lower mandibles. In the first year, the imm.'s head is darker grey than the ad.'s but it may become almost white in the second and third year, attaining the grey head of the adult in its fifth year. Imm. is best separated from imm. Blackbrowed by the all-dark bill. *Habitat.* Southern oceans, breeding during summer on sub-Antarctic islands. Rare north of 40° S and a vagrant to our coasts during winter. *Call.* Grunts and squawks when fighting over food. *Afrikaans.* Gryskopmalmok.

14 Yellownosed Albatross *Diomedea chlororhynchos* L 80 cm
The smallest and most slender of the 'black-backed albatrosses'. Ad. has a slight grey wash on head and nape but differs from ad. Greyheaded Albatross in that the yellow on the bill is confined to the ridge. Imm. differs from all other imm. albatrosses by its all-black bill and white head. Underwing in ad. and imm. differs from Blackbrowed and Greyheaded albatrosses by having narrower black borders. *Habitat.* Southern oceans, breeding in summer on sub-Antarctic islands. Common in all coastal waters, chiefly during winter. Most abundant off the Natal coast. *Call.* Loud bill clapping and a throaty 'weeeek'. *Afrikaans.* Geelneusmalmok.

34

11▼ 12▼   10▲

13▼ 14▼

15 Darkmantled Sooty Albatross *Phoebetria fusca* L 86 cm
An all-dark, very slender albatross with long narrow wings and an elongated, wedge-shaped tail which is usually held closed and appears pointed. The back is uniformly dark, matching the upperwings, and is never as pale as in Lightmantled Sooty Albatross. At close range the dark bill shows a cream to yellow stripe on the lower mandible and an incomplete white ring around the eye. Imm. of both *Phoebetria* albatrosses are difficult to tell apart but this species has a conspicuous buff collar and mottling on the back. Giant Petrels have a short rounded tail, much shorter wings and a heavy-bodied appearance. *Habitat.* Southern oceans, seldom north of 40° S in our sector. A rare vagrant to our coasts. *Call.* Silent at sea. *Afrikaans.* Bruinmalmok.

16 Lightmantled Sooty Albatross *Phoebetria palpebrata* L 86 cm
Differs from Darkmantled Sooty Albatross by having an ashy grey mantle which contrasts with darker head, wings and tail. The dark bill has a pale blue stripe on the lower mandible. Imm. is paler than imm. Darkmantled Sooty and the mottling on the back extends on to the lower back and rump.
Habitat. Southern oceans. A more southerly distribution than Darkmantled Sooty, rarely straying north of 45° S. A rare vagrant, with fewer than five records from our coasts. *Call.* Silent at sea . *Afrikaans.* Swartkopmalmok.

17 Southern Giant Petrel *Macronectes giganteus* L 90 cm
This species has an all-white phase which is unknown in the Northern Giant Petrel, otherwise these two species are very similar and can only be identified with certainty at close range. They are both large, the size of a small albatross, with heavy, pale-coloured bills. Flight is less graceful than that of the albatrosses: a heavy-bodied, stiff-winged, clumsy flapping motion. *Habitat.* Southern oceans. Common winter visitor to coastal waters. Scavenges around seal colonies in the Cape and Namibia. *Call.* Harsh grunts when squabbling over food. *Afrikaans.* Reuse Nellie.

18 Northern Giant Petrel *Macronectes halli* L 90 cm
Usually distinguished by a flesh-coloured bill, tinged with green, and a reddish brown tip. Southern Giant Petrel's bill is also a fleshy green colour but it has a green tip. The dark plumage is very variable in both species but this one tends more to grey. *Habitat.* Southern oceans. Occurs more frequently than the Southern Giant Petrel at seal colonies. Common winter visitor, chiefly to the west coast. Few seen during summer. *Call.* Similar to Southern Giant Petrel but slightly higher pitched. *Afrikaans.* Grootnellie.

20 Antarctic Petrel *Thalassoica antarctica* L 44 cm
Vaguely resembles the Pintado Petrel but lacks the chequered back of that species and shows a conspicuous white stripe in each wing. The head is dark and the white tail is narrowly tipped black. Underparts white. During the 24-hour sunlight of the Antarctic summer, the dark brown coloration bleaches to pale brown. *Habitat.* Antarctica, rarely north of the pack ice. One record for our region. *Call.* Usually silent at sea and at the pack ice. *Afrikaans.* Antarktiese Stormvoël.

21 Pintado Petrel *Daption capense* L 40 cm
A medium-sized black and white petrel with chequering on the back and two circular white patches on each upperwing. The head is black and the tail white with a black tip. Throat is grizzled and the underparts are white. Underwing is white, narrowly bordered with black. *Habitat.* Southern oceans, breeding on islands in the Antarctic and sub-Antarctic. Common winter visitor to our coasts, being extremely abundant on the trawling grounds of the western Cape. *Call.* When feeding, utters a high-pitched 'cheecheecheechee'. *Afrikaans.* Seeduifstormvoël.

5▲   17▼ 18▼   16◢

20▼ 21▼

23 Greatwinged Petrel *Pterodroma macroptera* L 42 cm
A dark brown petrel which differs from similarly sized Sooty Shearwater by having a dark, not silvery underwing and from the much larger Whitechinned Petrel by having a short black bill. In flight, the wings are held at a sharp angle at the wrist and the head appears heavy and downward pointing. Flight action in strong winds is fast and dynamic, twisting, turning and wheeling in high arcs over the sea. *Habitat.* Southern oceans, breeding on sub-Antarctic islands. Common visitor to the Cape and Natal coasts throughout the year but chiefly in late winter and early spring. *Call.* Silent at sea. A harsh, piercing 'keea-kee-kee-kee' is uttered over breeding grounds at night. *Afrikaans.* Langvlerkstormvoël.

27 Kerguelen Petrel *Pterodroma brevirostris* L 34 cm
Very similar to Greatwinged Petrel in outline shape but is smaller and is greyish, not dark brown. The head appears unusually large in this species and the sharp angle of the forehead shadows the face, imparting a hooded appearance in some lights. The dark underwings have pale areas on the leading edge and centre which reflect light and appear silvery. In a manner unusual for a petrel, this species flies very high over the ocean and sometimes hangs into the wind, much like a huge swift. *Habitat.* Southern oceans, breeding on sub-Antarctic islands in the summer. Very rare winter visitor to the Natal and Cape coasts, with most records being beached victims. *Call.* Silent at sea. *Afrikaans.* Kerguelense Stormvoël.

32 Whitechinned Petrel *Procellaria aequinoctialis* L 54 cm
Differs from all other dark brown petrels by its large size and long, robust, pale greenish bill. The white chin varies in extent: it may encompass the whole throat and cheeks or may even be absent. At long range resembles Fleshfooted Shearwater but has narrower, more pointed wings and is darker brown in colour. *Habitat.* Southern oceans, breeding on sub-Antarctic islands. A common visitor to all our coasts, chiefly in winter. Exceptionally abundant over the trawling grounds of the western Cape. *Call.* Normally silent at sea but if squabbling over food will utter a screaming 'tititititititi'. *Afrikaans.* Bassiaan.

36 Fleshfooted Shearwater *Puffinus carneipes* L 49 cm
A dark brown shearwater with a dark-tipped, flesh-coloured bill and flesh-coloured legs and feet. Its larger size and dark underwing separate it from Sooty Shearwater, while its larger size, its bill and leg colour, and rounded tail distinguish it from Wedgetailed Shearwater. At long range, if the feet are extended they appear as a pale-coloured area on the vent. *Habitat.* Coastal waters of the eastern Cape and Natal. Breeds on islands off Australia and arrives here in winter to follow the sardine run along the east coast. Uncommon. *Call.* Silent at sea. *Afrikaans.* Bruinpylstormvoël.

37 Sooty Shearwater *Puffinus griseus* L 46 cm
Differentiated from all other dark shearwaters by the silvery lining to the underwings. Wing beats are rapid, interspersed with short glides. *Habitat.* Most abundant in Cape coastal waters where it forages on small shoaling fish. Present throughout the year although more common in early spring. *Call.* Silent at sea. *Afrikaans.* Malbaatjie.

41 Wedgetailed Shearwater *Puffinus pacificus* L 45 cm
Similar to Fleshfooted Shearwater in colour and size but lacks that species' bright pink bill and feet. The tail is wedge-shaped but appears pointed in flight. The Sooty Shearwater is larger, has a silvery underwing and lacks the pointed tail. *Habitat.* Tropical and subtropical oceans. In southern African waters, a rare vagrant to Natal and the eastern Cape. *Call.* Silent at sea. *Afrikaans.* Keilstertpylstormvoël.

33▲   32▼   36▼   27▲

37▼   41▼

24 **Softplumaged Petrel** *Pterodroma mollis* L 35 cm
Resembles Atlantic Petrel but is much smaller and has a white throat and
forehead. Dark brown smudges on sides of breast sometimes form a complete
breast band. The underwing and tail are dark. Upperparts are dark with a faint,
open 'M' pattern across the upperwing. The rare dark phase resembles
Kerguelen Petrel but has broader, more rounded wings and is mottled on the
belly. Does not have the same high flying action as Kerguelen Petrel.
Habitat. Southern oceans, breeding on sub-Antarctic islands. Small
population breeds on islands of the North Atlantic but does not visit our
shores. The most common 'gadfly' petrel on the Cape and Natal coasts,
chiefly in winter. *Call.* Silent at sea. *Afrikaans.* Donsveerstormvoël.

25 **Whiteheaded Petrel** *Pterodroma lessonii* L 46 cm
The white head has distinctive black lozenge patches around the eyes.
Underparts white with contrasting dark grey underwings. Distinguished from
the similarly patterned, larger Grey Shearwater by its white head and undertail.
Upperparts greyish brown with a faint, open 'M' pattern across the wings. An
easily identified petrel with a fast, high-arcing flight action. Rarely attracted to
ships. *Habitat.* Southern oceans, not breeding near our sector. Rarely ventures
north of 40° S. An uncommon winter visitor to the Cape and Natal coasts.
Call. Silent at sea. *Afrikaans.* Witkopstormvoël.

26 **Atlantic Petrel** *Pterodroma incerta* L 44 cm
Similar in size and shape to Whiteheaded Petrel, this is a dark brown bird with
a conspicuous white lower breast and belly. In worn plumage, the breast and
throat can appear mottled brown but never pure white. The upperparts are a
uniform dark chocolate brown with no suggestion of an 'M' on the
upperwings. *Habitat.* Southern Atlantic Ocean where it is an endemic breeder
on the Tristan archipelago. Extremely rare in our region with most records from
the western Cape. Only one record from Natal. *Call.* Silent at sea.
Afrikaans. Bruinvlerkstormvoël.

28 **Blue Petrel** *Halobaena caerulea* L 30 cm
A small blue-grey petrel with white underparts and underwings. Diagnostic
features are the black markings on the crown and nape, and the square-
ended, white-tipped tail. The open 'M' pattern on the upperwings is less
distinct than on the prions. This species flies faster and arcs higher over the
waves than the prions. *Habitat.* Southern oceans, seldom straying above
40° S. A rare winter visitor to southern African waters, with most records being
beach-wrecked victims. *Call.* Silent at sea. A soft, dove-like cooing is given
from the nest burrow at night. *Afrikaans.* Bloustormvoël.

19 **Antarctic Fulmar** *Fulmarus glacialoides* L 48 cm
A very pale grey, gull-sized petrel, with white underparts and white wing
flashes at the base of the primaries. At close range the pink bill with its dark tip
is diagnostic. Unlikely to be confused with any other petrel in the region.
Habitat. Southern oceans, breeding in the far south on Antarctic islands and
mainland. Uncommon winter visitor to the Natal and Cape coasts. Most
frequently seen in the region of western Cape trawling grounds where it
scavenges offal. *Call.* Utters a high-pitched cackle when squabbling over
food. *Afrikaans.* Silwerstormvoël.

40

19 ▲

24 ▲   25 ▼

26 ▲   28 ▼

29 Broadbilled Prion *Pachyptila vittata* L 30 cm
Taxonomic research has shown that this species includes the Dove Prion *P. desolata* and Salvin's Prion *P. salvini*. The largest of the prions, it is identifiable at sea only if seen at very close range when the broad, flattened bill is diagnostic. Other characters to look for are the unusually large head and broad grey smudges on the sides of the breast. *Habitat.* Southern oceans, breeding on the Tristan and Prince Edward islands. The most abundant prion along our coasts, chiefly in winter. Subject to 'wrecks', when thousands of corpses are washed ashore. *Call.* Silent at sea. A loud, raspy dove-like cooing is given from the nest chamber at night. *Afrikaans.* Breëbekwalvisvoël.

30 Slenderbilled Prion *Pachyptila belcheri* L 27 cm
Distinguished with difficulty from Broadbilled Prion but is generally much paler, especially around the head, and lacks the broad grey smudges on the sides of the breast. When seen at close quarters the very thin bill is diagnostic. *Habitat.* Southern oceans, with a far more southerly distribution than other prions. Uncommon winter visitor to Cape and Natal waters. *Call.* Silent at sea. *Afrikaans.* Dunbekwalvisvoël.

31 Fairy Prion *Pachyptila turtur* L 24 cm
The smallest prion in the region and the one most easily identified. In coloration more blue, less grey than other prions with a very pale head and a diagnostic broad black tip to the tail. All other prions have narrow black tail bands. *Habitat.* Southern oceans, breeding on the Prince Edward group of islands. A rare winter visitor to Cape and Natal coasts. *Call.* Silent at sea. *Afrikaans.* Swartstertwalvisvoël.

22 Bulwer's Petrel *Bulweria bulwerii* L 27 cm
A small, prion-sized petrel which is uniformly dark brown and has a diagnostic long, wedge-shaped tail which is usually held closed and appears pointed. A pale brown stripe runs across each upperwing on the edges of the secondary coverts. Flight is buoyant and graceful, with the bird swooping over wave tops, then dipping into the next trough. *Habitat.* Open ocean, seldom close to shore. Breeds on islands in the North Atlantic. A very rare vagrant to the western Cape. *Call.* Silent at sea. *Afrikaans.* Bulwerse Stormvoël.

39 Little Shearwater *Puffinus assimilis* L 27 cm
This and the Audubon's Shearwater are the smallest shearwaters of the region. Very similar to Manx Shearwater but distinguished by its smaller size, dumpier shape, shorter wings and its flight action of very rapid wing beats punctuated by short glides. At close range the black cap ending above the eye and giving the bird a white-faced appearance, can be seen. Sub-Antarctic race has dark face but is dark grey, not black above. *Habitat.* Open oceans in mixed flocks of Sooty Shearwaters and Cape Gannets. Very rare visitor, chiefly in winter, to the western Cape coast. *Call.* Silent at sea. *Afrikaans.* Kleinpylstormvoël.

40 Audubon's Shearwater *Puffinus lherminieri* L 28 cm
In the field difficult to distinguish from Little Shearwater. When seen in good light, this species appears brown and white (not black and white as in Little Shearwater) and has dark, not white undertail coverts. The sub-Antarctic race of the Little Shearwater is dark grey above and can appear brownish: it is doubtful if this race and Audubon's Shearwater could be separated reliably in the field unless the undertail character is seen. *Habitat.* Tropical and subtropical oceans. In southern African waters, a rare vagrant to Natal, especially in summer. *Call.* Silent at sea. *Afrikaans.* Swartkroonpylstormvoël.

29▲ 31▼ 22▼ 30▲

39▼ 40▼

33 Grey Shearwater (Grey Petrel) *Procellaria cinerea* L 48 cm
This brown and white petrel resembles Cory's Shearwater from which it is
easily distinguished by having a dark underwing and tail. The grey-brown of
the head extends very low on the cheeks, leaving only a narrow white throat
patch, with the result that at long range this species can appear dark headed.
In flight, has shallow wing beats and a more direct flight action than other
petrels, with less banking and shearing on very stiff, straight wings.
Habitat. Southern oceans, breeding on the Tristan and Prince Edward groups
of islands. A rare winter visitor to the southern and western Cape coasts.
Call. Silent at sea. Display call a nasal 'ped-i-unker', followed by a fast rattling
trill. *Afrikaans.* Pediunker.

34 Cory's Shearwater *Calonectris diomedea* L 45 cm
The ashy brown upperparts and lack of both the dark-capped effect and pale
collar distinguish this large, heavy-bodied species from the Great Shearwater.
At closer range shows a yellow bill, and individuals display varying amounts of
white on the rump. Has a slower, more laboured flight than Great Shearwater
and flies lower over the sea, not shearing and banking as much as other
shearwaters. *Habitat.* Open oceans in both the North and South Atlantic.
Forages close inshore on small schooling fish. During summer, a common
European visitor to our coasts, except off Natal and Moçambique. *Call.* Silent
at sea. *Afrikaans.* Geelbekpylstormvoël.

35 Great Shearwater *Puffinus gravis* L 46 cm
Similar in size to Cory's Shearwater but distinguished by its distinctive black
cap, barred brown and buff back, dark smudge on lower belly and less
marked white underwing. Flight action is more dynamic than Cory's
Shearwater, flying higher over waves with more rapid wing beats.
Habitat. Open ocean, seldom close inshore, in North and South Atlantic.
Endemic breeder on Tristan archipelago. Common in Cape waters in early
and late summer. *Call.* Silent at sea. *Afrikaans.* Grootpylstormvoël.

38 Manx Shearwater *Puffinus puffinus* L 36 cm
After the Little and Audubon's, the smallest shearwater of the region. It is all
black above and white below with white underwings. Larger than the Little
Shearwater, it has a more extensive black cap which extends below the eye,
and a less fluttering flight. Flight action consists of long glides interspersed
with a series of rapid wing beats. *Habitat.* Open seas in North and South
Atlantic. Mainly coastal in the western Cape, associating with flocks of Sooty
Shearwaters. Rare visitor, chiefly in summer, from its breeding islands in
Europe. *Call.* Silent at sea. *Afrikaans.* Swartbekpylstormvoël.

33 ▲

35 ▼

34 ▲

38 ▼

42 European Storm Petrel *Hydrobates pelagicus* L 15 cm
The smallest seabird in our region. It commonly occurs in large flocks and the black plumage, white rump and white flash on the underwing help identify it. The feet do not project beyond the end of the tail in flight, which is bat-like and direct. The bird sometimes hovers, pattering its feet over the water when feeding. *Habitat.* Coastal waters of the North and South Atlantic. During summer, a common European visitor to the Cape and Natal coasts. *Call.* Silent in our region. *Afrikaans.* Swartpootstormswael.

43 Leach's Storm Petrel *Oceanodroma leucorhoa* L 20 cm
Larger than European Storm Petrel. Best identified by long narrow wings and distinctive flight action: it bounds low over the waves with quick jerky movements and sudden directional changes. When the bird spreads its tail, a clear fork can be seen. *Habitat.* Open oceans in the North and South Atlantic. Uncommon summer visitor from Europe to the Cape and Natal coasts. *Call.* Silent in our region. *Afrikaans.* Swaelstertstormswael.

44 Wilson's Storm Petrel *Oceanites oceanicus* L 18 cm
Differs from all other storm petrels in wing shape and flight action. Larger than European Storm Petrel, with broader, more rounded and flattened wings and long spindly legs, usually held projecting beyond end of tail. Yellow, webbed feet cannot be seen when the bird is in flight. Flight is swallow-like and action is direct with frequent glides. Long legs dangle in water when feeding. *Habitat.* Open oceans, breeding in Antarctica and on sub-Antarctic islands. Abundant visitor to all our coasts throughout the year but chiefly during winter. *Call.* Silent at sea. *Afrikaans.* Geelpootstormswael.

45 Whitebellied Storm Petrel *Fregetta grallaria* L 20 cm
Very difficult to distinguish from Blackbellied Storm Petrel but always appears much paler, has the back feathers edged with grey and the totally white belly extending on to the vent and undertail coverts. Accompanies ships, where it is seen more often in the bow waves than in the wake. *Habitat.* Southern oceans, with a more northerly distribution than Blackbellied Storm Petrel. Very rarely seen off the Cape and Natal coasts. *Call.* Silent at sea.
Afrikaans. Witpensstormswael.

46 Blackbellied Storm Petrel *Fregetta tropica* L 20 cm
Larger than Wilson's Storm Petrel from which it differs by its white underwings and white belly with a broad black line down the centre. Distinguished from Whitebellied Storm Petrel by appearing much darker and by the black line down its belly. Both species share similar flight characteristics, bounding over the waves and, when feeding, appearing to bounce off waves on their breasts. *Habitat.* Southern oceans. An uncommon winter and early summer visitor to the Cape and Natal coasts. *Call.* Silent at sea.
Afrikaans. Swartstreepstormswael.

46

42▲   43▲

46▲   44▼   45▼

47 Redtailed Tropicbird *Phaethon rubricauda* L 50 cm
Ad. is almost pure white (tinged pink when breeding) with a stout red bill and two extremely long, wispy, red central tail feathers. The tail feathers are not easily visible when the bird is seen at long range. Imm. differs from imm. Whitetailed Tropicbird by being more barred on the back, by having a greater expanse of black on the wings and by its black, not yellow bill. *Habitat.* Tropical oceans and Indian Ocean islands where it breeds. During summer, a rare visitor to our eastern shores. *Call.* Various breathy grunts and cackles. *Afrikaans.* Rooipylstert.

48 Whitetailed Tropicbird *Phaethon lepturus* L 44 cm
Ad. differs from Redtailed Tropicbird by having two black patches on each wing and two elongated white central tail feathers which are conspicuous even at long range. The bill is orange-yellow in both ad. and imm. (not black as in imm. Redtailed Tropicbird). *Habitat.* Tropical oceans and islands. Vagrant on east coast to Natal, with fewer than five records. *Call.* Silent in our region. *Afrikaans.* Witpylstert.

53 Cape Gannet *Morus capensis* L 85 cm
Ad. is white with black flight feathers and a black, pointed tail. The bill is heavy, long, pointed and pale grey, while the nape and sides of neck are straw yellow. Young imm. is dark brown with a paler, buff-speckled belly, but with age the plumage whitens and progresses through various mottled brown and white stages. In these plumages this species could be confused with the Brown Booby which is smaller and always shows a brown bib and white belly. *Habitat.* Coastal, not ranging very far out to sea. Breeds on islands off the Cape and Namibia. Common along all our shores. *Call.* When feeding in flocks at sea, the birds give a 'warrra-warrra-warrra' call. *Afrikaans.* Witmalgas.

54 Australian Gannet *Morus serrator* L 85 cm
Almost identical to the Cape Gannet but has a higher pitched call and a gular stripe only one third the length of that found in Cape Gannet. *Habitat.* As for Cape Gannet. One record only, that of a bird caught in a Cape Gannet colony in the western Cape. *Call.* Similar to Cape Gannet but higher pitched. *Afrikaans.* Australiese Malgas.

52 Brown Booby *Sula leucogaster* L 70 cm
Ad. is brown with a white belly and underwings. Similar to an imm. Cape Gannet, it has a long, cigar-shaped body outline, and a pointed bill and tail. Imm. is a darker version of ad. and lacks the white speckling of imm. Cape Gannet. *Habitat.* Tropical oceans and islands. Recorded from Moçambique and Natal. *Call.* Usually silent at sea. *Afrikaans.* Bruinmalgas.

51 Masked Booby *Sula dactylatra* ☐
No positive records of this species exist within our region but it is likely to occur off the Moçambique coast after cyclonic conditions. *Afrikaans.* Brilmalgas.

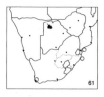

61 Greater Frigatebird *Fregata minor* L 95 cm
A black bird, larger than a Cape Gannet, with very long pointed wings. The elongated forked tail appears pointed when held closed. Imm. has whitish or tawny head and pale breast. *Habitat.* Tropical seas and islands. Recorded on east coast during or after cyclonic conditions. *Call.* Silent at sea. *Afrikaans.* Fregatvoël.

47▲   53▼ 54▼   48▲

                                 52▼ 61▼

55 Whitebreasted Cormorant *Phalacrocorax carbo* L 90 cm

The largest cormorant of the region. Ad. is glossy black with a white throat and breast, and a bright yellow patch at the base of the bill. During the breeding season white flank patches become evident. Imm. is dark brown with white underparts. *Habitat.* Both coastal and freshwater areas throughout the region. Breeds in trees around fresh water and on rocky islands in the western Cape. *Call.* Usually silent except during the breeding season when grunts and squeals are uttered. *Afrikaans.* Witborsduiker.

56 Cape Cormorant *Phalacrocorax capensis* L 65 cm

Intermediate in size between Bank and Crowned cormorants. Ad. has glossy blue-black plumage with a bright yellow-orange patch at base of bill. The patch brightens during the breeding season. Imm. is dark brown with slightly paler underparts. *Habitat.* A marine cormorant which rarely enters harbours. Long lines or 'treks' of flying birds can be seen just off the western Cape coast. Mostly in the southern and western Cape, Namibia, and occasionally north to Natal and Moçambique. *Call.* Silent except during the breeding season when various 'gaaaa' and 'geeee' noises are uttered. *Afrikaans.* Trekduiker.

57 Bank Cormorant *Phalacrocorax neglectus* L 76 cm

Larger and more robust than Cape Cormorant from which it differs by having dull black plumage and a thick woolly-textured neck, and by lacking the pale patch at base of bill. In breeding plumage shows white flecks on head and, for a short period, a diagnostic white rump. A small tuft of feathers on the forehead is erectile, giving the appearance of a small, rounded crest. *Habitat.* Coastal waters of the western Cape, and Namibian islands. Common in small parties on offshore islands in Benguela current. *Call.* Normally silent except in colonies. Call is a wheezy 'wheeee' given when bird alights near nest. *Afrikaans.* Bankduiker.

58 Reed Cormorant *Phalacrocorax africanus* L 52 cm

A small black cormorant with pale barring on back and a long, graduated tail. In breeding plumage ad. has a yellow-orange face patch and throat, and a small crest on forehead. Imm. is dark brown above and white below. A freshwater cormorant rarely seen at the coast. *Habitat.* Common throughout the region on freshwater dams, lakes and slow-moving streams. Avoids saltwater estuaries and bays, except in coastal Moçambique. *Call.* Silent except when breeding. *Afrikaans.* Rietduiker.

59 Crowned Cormorant *Phalacrocorax coronatus* L 50 cm

Very like Reed Cormorant but differs by having a much shorter tail and by having an orange-red face and throat patch when in breeding dress. The erectile crest on the forehead is longer and more pronounced than that of Reed Cormorant. Imm. is dark brown above and differs from imm. Reed Cormorant by its shorter tail and brown, not white underparts. *Habitat.* Almost entirely marine, on islands off the western Cape and Namibian coasts. May be seen in estuaries and lagoons in the western Cape. *Call.* Normally silent except when breeding. *Afrikaans.* Kuifkopduiker.

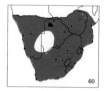

60 Darter *Anhinga melanogaster* L 80 cm

The long egret-like neck, combined with slender head and elongated pointed bill should rule out confusion with cormorants. Ad. in breeding plumage shows rust-coloured head and neck with long white stripe running from eye on to neck. Non-breeding bird is pale brown on face and throat. When swimming, sometimes totally submerges body, leaving only its long neck and head showing, scything through the water with serpentine movements. *Habitat.* Common on freshwater dams, lakes and slow-moving streams throughout the region. Sometimes found in coastal lagoons and estuaries. *Call.* Normally silent. *Afrikaans.* Slanghalsvoël.

57 ▲   56 ▲   58 ▲   60 ▼

55 ▲   59 ▼

62 Grey Heron *Ardea cinerea* L 100 cm
A large, greyish heron distinguished from others by its white head and neck, and a black eye-stripe which ends on the nape in a wispy black plume. Bill is long, dagger shaped and yellow. Imm. lacks eye-stripe and has darker bill. In flight differentiated from Blackheaded Heron by uniform grey underwings. *Habitat.* Frequents a wide range of fresh- and salt-water areas. Common throughout the region except in deserts. *Call.* A harsh, booming 'kraaunk' given in flight. *Afrikaans.* Bloureier.

63 Blackheaded Heron *Ardea melanocephala* L 96 cm
Slightly smaller than Grey Heron. The black-topped head and hind neck contrast with white throat. In flight the white wing linings highlight the dark flight feathers. Imm. has grey, not black on head. *Habitat.* More often seen stalking through open grasslands than around water. Common throughout the region except in deserts. *Call.* A loud 'aaaaark', and various hoarse cackles and bill clappering at nest. *Afrikaans.* Swartkopreier.

64 Goliath Heron *Ardea goliath* L 140 cm
Its large size should be diagnostic. Most closely resembles the Purple Heron, but the latter has a black crown and black throat stripes and is very much smaller. At long range can be distinguished from Grey Heron by its rufous head and neck. *Habitat.* Freshwater dams, lakes and sometimes estuaries. Thinly distributed in the north and east, but absent from the west and south. Usually solitary or in pairs. *Call.* A loud, low-pitched 'kwaaark'. *Afrikaans.* Reuse Reier.

65 Purple Heron *Ardea purpurea* L 91 cm
Noticeably smaller and slimmer in appearance than the more common Grey Heron. In coloration most closely resembles the Goliath Heron but has a black, not rufous crown and is almost half that species' size. Imm. lacks the grey nape and mantle of ad., is less streaked on the neck and has far less black on the crown. *Habitat.* Thick stands of reeds: reluctant to feed in the open. Common in suitable habitat except in the dry central and western regions. *Call.* Similar to Grey Heron's 'kraaark'. *Afrikaans.* Rooireier.

50 Pinkbacked Pelican *Pelecanus rufescens* L 140 cm
Considerably smaller than the White Pelican, this species differs by being dowdier and greyer, and by the forehead feathers meeting the bill in a straight line. In flight the primaries and secondaries do not contrast with the coverts and appear almost uniform. Imm. is dark brown at first and whitens progressively with age. Best distinguished from White Pelican by size and bill feather character. *Habitat.* Coastal estuaries, less often on fresh water. Common on east coast as far south as Natal. Vagrant to the Cape. *Call.* Usually silent except when breeding. *Afrikaans.* Kleinpelikaan.

49 White Pelican *Pelecanus onocrotalus* L 180 cm
A very large white bird which assumes a pinkish flush in the breeding season. The white forehead feathers project on to the bill as a pointed wedge but this is really only noticeable at close range. Black primaries and secondaries contrast with white coverts. Imm. is dark brown and whitens progressively with age. *Habitat.* Groups habitually fish on open freshwater lakes. Frequents estuaries in the western Cape, Namibia and Natal. *Call.* Usually silent except when breeding. *Afrikaans.* Witpelikaan.

62▲
64▼
65▼
63▲
60▼
49▼

66 Great White Egret *Egretta alba* L 95 cm

The largest white heron of the region. Breeding bird has black bill with a yellow base but out of the breeding season the whole bill turns yellow. Differs from Yellowbilled Egret by its larger size, longer and heavier bill with gape extending behind the eye, and much longer, thinner neck, usually held kinked in an 'S' shape. Distinguished from Little Egret by much larger size and black, not yellow toes. *Habitat.* Freshwater dams, lakes, flooded meadows, and marine estuaries and lagoons. Common throughout the region except in the dry west. *Call.* A low, heron-like 'waaaark'. *Afrikaans.* Grootwitreier.

67 Little Egret *Egretta garzetta* L 65 cm

When seen, the yellow toes distinguish this species from any other white heron in our region. The bill is slender and always black. Feeds characteristically by dashing to and fro in shallow water in pursuit of prey. Usually forages alone: not gregarious except when breeding. *Habitat.* Most freshwater situations, both natural and man-made, and frequently along rocky coastlines and estuaries. Common throughout the region in suitable habitats. *Call.* Similar to other egrets, a harsh 'waaark'. *Afrikaans.* Kleinwitreier.

68 Yellowbilled Egret *Egretta intermedia* L 66 cm

Intermediate in size between Cattle and Great White egrets. Separated from Great White Egret by noticeably shorter, thicker neck which is not held in such a severe 'S' shape. Gape does not extend behind the eye but ends just below it. Although not easily seen, yellowish green garters are diagnostic. Differs from Cattle Egret by larger size, longer bill and more slender appearance. *Habitat.* Flooded veld and marshes and any damp grassy areas, but infrequently near open water. Common throughout the region except in desert areas. *Call.* Typical heron-like 'waaaark'. *Afrikaans.* Geelbekwitreier.

71 Cattle Egret *Bubulcus ibis* L 54 cm

Breeding birds have red bill, and buff plumes on head, breast and mantle, but they never become as dark as the Squacco Heron. Bill is shorter and more robust than in other white herons and there is a noticeable shaggy bib and throat which give this species a distinct jowl. The legs are never black, but vary from dark brown through to yellowish green, and are red during the breeding season. *Habitat.* Much drier areas than other egrets and often found in association with cattle or game. Common throughout the region except in desert areas. *Call.* Typical heron-like 'aaaark' or 'pok-pok'. *Afrikaans.* Bosluisvoël (Veereier).

72 Squacco Heron *Ardeola ralloides* L 42 cm

The smallest white heron of the region. At rest, looks more bittern-like but when in flight shows all-white wings. Unlikely to be confused with Cattle Egret because of its very buffy appearance and its habits. *Habitat.* Heavily vegetated freshwater lakes and slow-moving rivers. Skulks in reedbeds and long grass, sitting motionless for hours. Common in suitable habitat in the north and east. *Call.* Low-pitched, rattling 'kek-kek-kek'. *Afrikaans.* Ralreier.

73 Madagascar Squacco Heron *Ardeola idae* L 47 cm

Breeding plumage is completely white. In non-breeding dress very difficult to distinguish from Squacco Heron but is slightly larger, has a noticeably heavier bill and much broader streaking on the throat and breast. *Habitat.* Similar to Squacco Heron but is found more often in open situations. Only one record for the region, in the extreme north-east. Migrant from Madagascar, mainly to East Africa during winter. *Call.* Usually silent. *Afrikaans.* Malgassiese Ralreier.

67▲

66▼

68▼

72▲

73▼

71▲

70 Slaty Egret *Egretta vinaceigula* L 60 cm
Very similar to Little Egret in shape and behaviour. In colour resembles Black Egret but differs by having greenish yellow legs and feet, a rufous throat and by lacking the wing canopy feeding action. *Habitat.* Uncommon and confined to a small stretch of the Zambezi River and flooded grasslands and marshes in the Okavango swamps. *Call.* Unknown. *Afrikaans.* Rooikeelreier.

69 Black Egret *Egretta ardesiaca* L 66 cm
A small, slate black version of the Little Egret, with black legs and yellow toes. Imm. slightly paler than ad. Ad. lacks the rufous throat of the Slaty Egret. Diagnostic feeding behaviour of forming an 'umbrella' over the head with wings held forward and outstretched. *Habitat.* Freshwater lakes and dams, occasionally estuaries. Rare in the east and south, becoming more common in the north. Absent from the west. *Call.* Normally silent. *Afrikaans.* Swartreier.

75 Rufousbellied Heron *Ardeola rufiventris* L 58 cm
A small heron with a sooty black head and breast, and rufous belly, wings and tail. Female is duller version of male, and imm. has buff-streaked head and neck. In flight the bright yellow legs and feet contrast strongly with dark underparts. *Habitat.* Freshwater lakes and dams surrounded by tall reeds. Normally only seen when put to flight. Uncommon summer visitor in the east but resident further north. *Call.* Typical heron-like 'waaaaak' and other low grunts. *Afrikaans.* Rooipensreier.

79 Dwarf Bittern *Ixobrychus sturmii* L 25 cm
Dark slaty blue above, buff below overlaid with dark vertical stripes running from the throat down on to the breast. The tiniest heron in the region, appearing more like a large rail in the field. Imm. is similar to ad. but has the head and back streaked with buff. Distinguished from imm. Little Bittern by being generally darker and by being more heavily streaked below. *Habitat.* Freshwater ponds and lakes surrounded by trees, and especially in flooded woodlands. Occurs in the north and east. *Call.* Soft frog-like croaks. *Afrikaans.* Dwergrietreier.

78 Little Bittern *Ixobrychus minutus* L 36 cm
Differs from smaller Dwarf Bittern by having conspicuous whitish wing patches and less striping on throat and breast. Female is browner than male. Imm. resembles the female, but is much more heavily streaked below. The imm. is distinguished from Greenbacked Heron by smaller size and green, not orange legs. *Habitat.* Thick reedbeds. Uncommon in suitable habitat in the north and east, and in coastal south-western Cape. *Call.* Soft frog-like cheeping. *Afrikaans.* Woudapie (Kleinrietreier).

80 Bittern *Botaurus stellaris* L 64 cm
The largest bittern, it is more often heard than seen. The upperparts are mottled buff-brown and black with a dark crown and broad, conspicuous moustachial stripes. Underparts buff, heavily streaked dark brown. The black moustachial stripes should eliminate confusion with imm. Blackcrowned Night Heron. *Habitat.* Thick stands of reeds surrounding fresh water and large swamps. Status uncertain, as it is easily overlooked. *Call.* Deep, resonant booming, mostly at night. *Afrikaans.* Grootrietreier (Roerdomp).

74 Greenbacked Heron *Butorides striatus* L 40 cm
A small, dark grey heron with a black crown, dark green back and slightly paler underparts. The legs and feet are bright orange-yellow. The black, wispy nape plume is not usually seen except when bird alights. Imm. is streaked brown and buff, and has orange legs. *Habitat.* Frequents mangrove stands and coral reefs at low tide, and freshwater dams, lakes and sluggish rivers overhung with trees. Fairly common, but local and often overlooked. *Call.* A loud 'baaek' when alarmed. *Afrikaans.* Groenrugreier.

78 ▲   69 ▲   70 ▼   79 ▲   75 ▼

80 ▼   74 ▼

82 Shoebill *Balaeniceps rex* ☐
An unconfirmed report from the Okavango swamps is the only record in our region. *Afrikaans.* Skoenbekooievaar.

89 Marabou Stork *Leptoptilos crumeniferus* L 150 cm
The huge size, naked head and neck, combined with the massive bill render this species unmistakable. In flight the black wings contrast with the white body and the head is tucked into the shoulders. Imm. similar to ad. but has head and neck covered with a sparse woolly down. *Habitat.* Not often seen outside major game reserves where it scavenges at lion kills. Uncommon and thinly distributed in the north of the region. *Call.* A low, hoarse-sounding croak is given when alarmed. Clappers bill when displaying. *Afrikaans.* Maraboe.

87 Openbilled Stork *Anastomus lamelligerus* L 94 cm
A large black stork with a pale ivory-coloured bill which has a diagnostic wide, nutcracker-like gap between the mandibles. Imm. is duller version of ad. and lacks the bill gap as that only develops with maturity. *Habitat.* Freshwater lakes and dams. Summer visitor to the north and east. Rare except in the east. *Call.* Silent in our area. *Afrikaans.* Oopbekooievaar.

88 Saddlebilled Stork *Ephippiorhynchus senegalensis* L 145 cm
Unlikely to be confused with any other stork species because of its large size and its peculiar red and black banded bill, with the yellow 'saddle' at the bill base noticeable at close range. Imm. is grey instead of black, its neck and head are brown, and it lacks the yellow 'saddle' at the bill base. *Habitat.* Always close to fresh water or seen striding over nearby grasslands. Uncommon resident in the north and east of the region. *Call.* Normally silent except for bill clappering during display. *Afrikaans.* Saalbekooievaar.

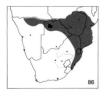

86 Woollynecked Stork *Ciconia episcopus* L 85 cm
The glossy black plumage, white woolly neck, and white belly and undertail are distinctive. The black tail is slightly forked and the legs and bill are dull reddish. Imm. is like ad. but the glossy black is replaced by brown and the black forehead is streaked white. *Habitat.* Usually near fresh water: lagoons, ponds and rivers. Uncommon and thinly distributed in the north and east. Solitary or in pairs. *Call.* Normally silent except when breeding. *Afrikaans.* Wolnekooievaar.

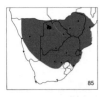

85 Abdim's Stork *Ciconia abdimii* L 76 cm
Distinguished from similar Black Stork by having a white lower back and rump, and greenish legs and bill. Imm. has dark red bill. In flight, the legs do not project as far beyond the end of the tail as in the Black Stork. *Habitat.* Open fields and agricultural lands, often in the company of White Storks. Common summer visitor, usually in large flocks. *Call.* Silent in our region. *Afrikaans.* Kleinswartooievaar (Blouwangooievaar).

84 Black Stork *Ciconia nigra* L 97 cm
A large, glossy black stork with a white belly and undertail. Distinguished from the smaller Abdim's Stork by the black rump and lower back, and by the red bill and legs. Imm. is a browner version of the ad. and has a yellowish bill. *Habitat.* Breeds on cliff ledges in mountainous areas and feeds in fast-flowing streams and open grasslands. Also occurs along the coast in estuaries and lagoons. Uncommon everywhere. *Call.* Silent except on nest when loud whining and bill clappering are given. *Afrikaans.* Grootswartooievaar.

58

87▲   89▼ 88▼   86▲

85▼ 84▼

83 White Stork *Ciconia ciconia* L 102 cm
Resembles the Yellowbilled Stork but differs by having a red bill and legs, and an all-white tail. Legs are often white because the birds excrete on them to lose body heat. Imm. has a darker bill and legs, and white plumage tinged with brown. *Habitat.* Open grassland and farming country. Common summer visitor throughout region except the dry west. A small breeding population is resident in the southern Cape. *Call.* Usually silent. *Afrikaans.* Witooievaar.

90 Yellowbilled Stork *Mycteria ibis* L 95 cm
The long, slightly decurved yellow bill is diagnostic. During the breeding season the naked facial skin takes on a reddish colour. In flight appears similar to White Stork but differs by having a black tail. Wing coverts and back are tinged with pink. Imm. similar to ad. but lacks the pink wash, and its head and neck are greyish, not white. *Habitat.* Lakes, large rivers and estuaries. Common resident in the north but rare summer visitor further south. *Call.* Normally silent except during breeding season when it gives loud squeaks and hisses. *Afrikaans.* Nimmersat.

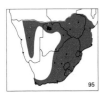

95 African Spoonbill *Platalea alba* L 90 cm
The long, flattened, spoon-shaped grey and red bill is diagnostic. In flight differs from similarly sized white egrets by holding its neck outstretched, not tucked into the shoulders. Imm. has a pinkish bill and dark-tipped flight feathers. Feeds with a characteristic side-to-side sweeping motion. *Habitat.* Freshwater lakes and estuaries. Thinly distributed in the east and north of the region. Absent from the more arid areas. *Call.* Normally silent except if alarmed, when it utters a low 'kaark'. In breeding colonies emits various grunts and clappers bill. *Afrikaans.* Lepelaar.

81 Hamerkop *Scopus umbretta* L 56 cm
A dark brown, long-legged bird with a long crest and flattened bill. The hammer-shaped profile of the head renders this bird unmistakable. In flight, its shape and barred tail lend a hawk-like appearance, but the long bill and legs rule out confusion with any bird of prey. *Habitat.* Freshwater dams, lakes and rivers. Common in suitable localities throughout the region. Absent from desert areas. *Call.* A jumbled mixture of hisses, squeaks and frog-like croaks. *Afrikaans.* Hamerkop.

96 Greater Flamingo *Phoenicopterus ruber* L 127 cm
A white bird with a small body in relation to its very long legs and neck. In flight shows brilliant red patches on the forewings. Differs from Lesser Flamingo in being larger, less red in colour and having a large, black-tipped pink bill. Imm. has white plumage tinged with brown and lacks the red in the wings. *Habitat.* Shallow freshwater lakes, salt pans and estuaries. Common where it occurs, sometimes in huge concentrations. *Call.* 'Honk, honk', sounding not unlike a farmyard goose. *Afrikaans.* Grootflamink.

97 Lesser Flamingo *Phoenicopterus minor* L 100 cm
Distinguished from Greater Flamingo by being much smaller and having a black-tipped, dark red bill which appears all black when seen at long range. Head, neck and body plumage very variable but normally far redder than Greater Flamingo, with a larger expanse of crimson in the wings. Imm. differs from imm. Greater Flamingo by being smaller, having a stubbier bill and greyish brown body coloration. *Habitat.* Freshwater lakes, salt pans and estuaries, usually in the company of the Greater Flamingo. Common throughout in suitable areas, being absent from the more arid regions. *Call.* A goose-like honking. *Afrikaans.* Kleinflamink.

60

83 ▲   95 ▼ 81 ▼   90 ▲

96 ▼   97 ▼

91 Sacred Ibis *Threskiornis aethiopicus* L 90 cm
A white bird with an unfeathered black head and neck, and a long, decurved black bill. The scapular feathers are blue-black and flight feathers are black tipped, giving a narrow black edge to the wing. During the breeding season, the naked skin on the underwing turns bright scarlet. Imm. differs from ad. by having the neck feathered white. *Habitat.* Coastal and on offshore islands in the western Cape. Found near fresh water, foraging in open areas. Common throughout the region except in more arid areas. *Call.* Loud croaking at breeding colonies. *Afrikaans.* Skoorsteenveër.

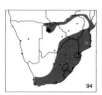

94 Hadeda Ibis *Bostrychia hagedash* L 76 cm
A drab, greyish brown bird which at close range shows glossy bronze wing coverts and back. The face is grey and a white stripe runs from the bill base to below and behind the eye. The long, dark, decurved bill has a red ridge on the upper mandible. Imm. lacks the red on bill and the bronzing on back and wings. *Habitat.* Breeds and roosts in woodlands, foraging in open glades and grasslands. Common in the eastern part of the region. Absent from the dry west. *Call.* One of the most familiar calls in Africa. Birds flying to and from roost utter a loud 'ha-ha-ha-dah-dah'. *Afrikaans.* Hadeda.

92 Bald Ibis *Geronticus calvus* L 78 cm
Differs from the smaller Glossy Ibis by having a bald red head with a white face, a long, decurved red bill and red legs and feet. General colour is glossy blue-black with coppery patches on forewings. Imm. has duller plumage with the head covered in short brown feathers and the red on the bill confined to the base. Legs are brown, not red as in ad. and the coppery patches on the forewings are absent. *Habitat.* Restricted to upland grasslands, especially burnt areas. Breeds on cliffs near waterfalls. Endemic. *Call.* During display utters a high-pitched 'keeaa-wop-wop'. *Afrikaans.* Kalkoenibis.

93 Glossy Ibis *Plegadis falcinellus* L 65 cm
The smallest ibis of the region. Appears black when seen at long range, but closer views show the head, neck and body to be a dark chestnut and the wings, back and tail a dark glossy green with bronze and purple highlights. Imm. has the glossy areas of ad. but the remainder of the body plumage is a dull, dark brown. *Habitat.* Freshwater lakes, swampy meadows and flooded grasslands. Uncommon in the eastern and northern sectors of the region. Small population breeds in the western Cape. *Call.* Normally silent. A low, guttural 'kok-kok-kok' given in breeding colonies. *Afrikaans.* Glansibis.

77 Whitebacked Night Heron *Gorsachius leuconotus* L 53 cm
The large dark head with conspicuous pale eye area, dark back and wings render this species unmistakable. Only in flight is the small white patch on the back visible. Imm. distinguished from imm. Blackcrowned Night Heron by having a darker crown and an unmarked face, and from Bittern by lack of black moustachial stripes. *Habitat.* Slow-moving streams overhung with thick tangles of reeds and trees. An easily overlooked, rare species confined to the east and north. *Call.* When disturbed gives a sharp 'kaaark'. *Afrikaans.* Witrugnagreier.

76 Blackcrowned Night Heron *Nycticorax nycticorax* L 56 cm
The black crown, nape and back contrast greatly with grey wings and tail, and white underparts. Imm. superficially resembles Bittern but is paler and lacks black crown and moustachial stripes. *Habitat.* Freshwater dams, streams and lagoons. During the day, flocks roost in reedbeds and trees, venturing out at dusk to feed. Common throughout the region except the dry west. *Call.* A harsh 'kwok-kwok' given when flushed from cover. *Afrikaans.* Gewone Nagreier.

91 ▲

93 ▼

94 ▲

92 ▼

76 ▼

77 ▲

116 Spurwinged Goose *Plectropterus gambensis* L 100 cm
A large black goose with variable amounts of white on the throat, neck and belly. Imm. resembles the ad. In flight the large white area on the forewing distinguishes this species from Knobbilled Duck. Feeds ashore on grasses and other vegetable matter. *Habitat.* Frequents water bordered by grasslands and agricultural areas. Common except in the south and dry west. *Call.* Utters a whistle in flight. *Afrikaans.* Wildemakou.

102 Egyptian Goose *Alopochen aegyptiacus* L 70 cm
A large, brown bird which has a dark brown mask over the eye, and a brown patch on the chest which is distinctive in flight. Compared with South African Shelduck it has a longer neck and legs, and a thin dark line running through the white forewing is visible in flight. Sexes alike. Imm. lacks the brown mask and chest patch and the forewing is not pure white. *Habitat.* Almost any freshwater habitat but frequently feeds in fields. Very common throughout. *Call.* Female utters a grunting honk, the male hisses. *Afrikaans.* Kolgans.

103 South African Shelduck *Tadorna cana* L 64 cm
A large, russet-coloured duck with black bill and legs. The male has a diagnostic grey head. Female has a variable white and grey head, whereas the Whitefaced Duck has a white and black head. In flight both sexes show white forewings but no black dividing line as seen in Egyptian Goose. *Habitat.* Freshwater lakes and dams in drier areas. Common endemic to the west and south. *Call.* Various honks and hisses. *Afrikaans.* Kopereend.

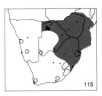

115 Knobbilled Duck *Sarkidiornis melanotos* L ♂ 80 cm; ♀ 65 cm
This large duck with its grey speckled head and contrasting blue-black and white plumage is unmistakable in the region. Male is larger than female and when breeding has a conspicuous comb on the bill. Imm. like female but has dark speckling on the white areas. In flight the wings are black with no markings although female shows a little white on the lower back. *Habitat.* Virtually any still, freshwater body. Widely distributed but nomadic, it is more common to the north. *Call.* Utters a whistle but is usually silent. *Afrikaans.* Kobbeleend.

99 Whitefaced Duck *Dendrocygna viduata* L 48 cm
A distinctive, long-necked duck with a diagnostic white face. Differs from larger South African Shelduck by lacking grey on the head. Like the Fulvous Duck, it stands very erect and is highly gregarious, but it can be distinguished from that species in flight by the lack of white on the rump. *Habitat.* Almost any expanse of water but spends much time ashore. Common and widespread except in the south. *Call.* A familiar three-note whistle, the last two notes being closer together. *Afrikaans.* Nonnetjie-eend.

100 Fulvous Duck *Dendrocygna bicolor* L 46 cm
Similar in shape to Whitefaced Duck but lacks the white face. Shows a dark line down the hind nape and neck and has conspicuous white flank stripes. In flight, distinguished from Whitefaced Duck by its white rump contrasting with the dark upperparts. Imm. distinguished from imm. Whitefaced Duck by white on flanks and rump. *Habitat.* Freshwater lakes and dams. Generally absent from the south and west. Nomadic, but generally is less abundant than Whitefaced Duck. *Call.* A soft, double-syllable whistle. *Afrikaans.* Fluiteend.

98 Mute Swan *Cygnus olor* L 102 cm
Unmistakable. A large, white bird with an orange-red bill and a black knob on the forehead. Imm. plumage is buffy, not white. *Habitat.* Coastal and inland bodies of fresh water. A feral population in the south has now apparently disappeared. Further sightings could be escapees from captivity. *Call.* Quiet but utters a nasal grunt when defending its nest or brood. *Afrikaans.* Swaan.

64

116▲

102▼

99▲

98▼

100▼

103▲

115▼

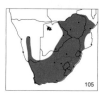

105 **African Black Duck** *Anas sparsa* L 56 cm
Resembles Yellowbilled Duck but is darker, has white spots on the back and a dark, not yellow bill. The feet are yellow. In flight, might be confused with Yellowbilled Duck as it also has a blue-green speculum edged with white, but in this species the wing linings are whitish, not grey as in the Yellowbilled. When sitting in the water it appears long bodied and sits low. *Habitat.* Prefers fast-moving streams and rivers. Widespread except in the dry west and desert areas. *Call.* 'Quack', especially in flight. *Afrikaans.* Swarteend.

104 **Yellowbilled Duck** *Anas undulata* L 54 cm
The rich yellow bill with a black patch on the upper mandible is distinctive. From a distance, appears dark but less so than African Black Duck. In flight it shows a blue-green speculum narrowly edged with white. *Habitat.* Forms mixed flocks on virtually any area of open water. Common, occurs throughout most of the region. *Call.* Female quacks. The male's call is a rasping hiss. *Afrikaans.* Geelbekeend.

109 **Pintail** *Anas acuta* L 47 cm
Male is striking with a dark chocolate brown head and a white stripe running down the side of the neck to the white breast. In eclipse plumage it resembles the female. Female is similar to Yellowbilled Duck but has a dark bill, paler plumage and a long, slender neck and body. The tail is pointed in both sexes, but the male's is longer in breeding plumage. *Habitat.* Inland water bodies. Recorded as far south as the Transvaal. Uncommon summer visitor. *Call.* Female quacks. Male's call is a soft nasal honk. *Afrikaans.* Pylsterteend.

108 **Redbilled Teal** *Anas erythrorhyncha* L 48 cm
The combination of dark cap, pale cheeks and red bill is diagnostic. Larger than Hottentot Teal which has a blue, not red bill. Dark cap distinguishes it from Cape Teal. In flight the secondaries are pale and the lack of the dark green speculum precludes confusion with Hottentot Teal. Sexes are similar. *Habitat.* Occurs in mixed flocks on fresh water. Very common throughout. *Call.* Female quacks. Male gives a soft, nasal whistle. *Afrikaans.* Rooibekeend.

107 **Hottentot Teal** *Anas hottentota* L 35 cm
A diminutive species resembling the Redbilled Teal from which it differs by having a blue, not red bill. In flight shows a green speculum with a white trailing edge and a black and white underwing. Sexes are similar. *Habitat.* Mainly inland on small water bodies, often close to floating vegetation. Uncommon in the south but occurs more widely to the north. *Call.* A soft whistle uttered in flight. *Afrikaans.* Gevlekte Eend.

110 **Garganey** *Anas querquedula* L 38 cm
The male has an obvious white superciliary stripe on a dark brown head. The brown breast is sharply demarcated by the white belly and the black and white stripes on the back are conspicuous. Male in eclipse plumage resembles female but has a more pronounced eyebrow stripe. Female is larger than Hottentot Teal and lacks the dark cap. In flight both sexes show the pale blue forewings also found in Cape and European shovelers, but lack their heavy bills. *Habitat.* Inland water bodies. A rare summer migrant to Zimbabwe and the Transvaal. *Call.* A nasal 'quack'. *Afrikaans.* Somereend.

106 **Cape Teal** *Anas capensis* L 46 cm
The palest duck of the region. At close range, the speckled plumage and pink bill are obvious. Distinguished from Redbilled Teal by the uniformly coloured head. In flight shows a dark greenish speculum surrounded by white. Sexes alike. Usually occurs in mixed flocks. *Habitat.* Any area of open fresh or saline water. Occurs throughout except in the north-east. *Call.* A thin whistle usually given in flight. *Afrikaans.* Teeleend.

05 ▲

108 ▼

104 ▲

109 ▼

107 ▼

110 ▲

106 ▼

111 **European Shoveler** *Anas clypeata* L 53 cm
Breeding plumage of male is unmistakable: green head, white chest and chestnut belly. In eclipse plumage the male resembles the female. Female differs from Cape Shoveler by being much paler overall, especially around the head and neck. In both sexes, European Shoveler has a heavier, longer and more spatulate bill than Cape Shoveler. Very rapid, direct flight shows powder blue forewing similar to the Garganey, but the latter is smaller and lacks the large bill. *Habitat.* Favours coastal water bodies but does occur inland. Uncommon migrant to Zimbabwe and the Transvaal. *Call.* Female quacks. Male gives a nasal 'crook, crook'. *Afrikaans.* Europese Slopeend.

112 **Cape Shoveler** *Anas smithii* L 53 cm
Both sexes distinguished from other ducks, except female European Shoveler, by the long, black, spatulate bill. The legs are a rich yellow-orange. The male has a paler head and yellower eyes than his female counterpart. Female is darker than European Shoveler, especially around the head and neck. *Habitat.* Fresh water, preferably with surface vegetation. Endemic, rarer in the north-east and abundant in the Cape and Transvaal. *Call.* 'Quack'. *Afrikaans.* Kaapse Slopeend.

117 **Maccoa Duck** *Oxyura maccoa* L 46 cm
The male has a chestnut body, large black head and a bright cobalt blue bill. (The male Southern Pochard has a smaller brown head and pale blue bill.) The female is dark brown with a pale cheek, under which is a dark line. This gives the impression of a pale stripe running from the bill beyond the eye: in comparison, the female Southern Pochard has a vertical crescent under the eye. Imm. and male in eclipse plumage resemble the female. In flight, the upperwing is uniform dark brown. Sits very low in the water with its stiff tail often cocked at a 45° angle. *Habitat.* A diving species of quieter waters, with surface vegetation. Thinly distributed throughout, absent from the desert regions and eastern lowlands. *Call.* A peculiar nasal trill. *Afrikaans.* Bloubekeend (Makou-eend).

113 **Southern Pochard** *Netta erythrophthalma* L 50 cm
Male is similar to male Maccoa Duck but has a blue, not cobalt-coloured bill, bright red eyes and a brown, not black head. The female is dark brown with a pale patch at the bill base and a pale crescent extending down from the eye. In flight both sexes appear brown with a distinct white wing bar. Sits low in the water with its tail submerged. *Habitat.* Lakes, dams and vleis. A common, sociable bird. *Call.* The male makes a whining sound, the female quacks. *Afrikaans.* Bruineend.

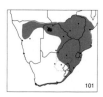

101 **Whitebacked Duck** *Thalassornis leuconotus* L 43 cm
A mottled brown, hump-backed, large-headed duck. When swimming the most diagnostic feature is the pale patch at the base of the bill. Only in flight is the white back seen. Male, female and imm. are similar. *Habitat.* Fresh water, often difficult to see in floating vegetation. Recorded throughout, except in the drier regions. *Call.* Normally silent but does whistle softly. *Afrikaans.* Witrugeend.

114 **Pygmy Goose** *Nettapus auritus* L 33 cm
The smallest of our geese and ducks. The orange body, white face and dark greenish upperparts are diagnostic. Male has highly contrasting head markings whereas the female's head tends to be more speckled. In flight, the large white wing patches show up distinctly. Difficult birds to see as they sit motionless in floating vegetation. *Habitat.* Prefers freshwater areas with floating vegetation. Recorded sporadically throughout but more common along the eastern coastal belt and in the north of the region. *Call.* A soft whistle. *Afrikaans.* Dwerggans.

68

111▲   112▼   117▼   113▲

101▼   114▼

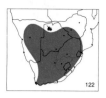

122 Cape Vulture *Gyps coprotheres* L 115 cm
Difficult to distinguish from ad. and imm. Whitebacked Vultures. When seen together, the Cape Vulture is much larger and is usually much whiter in appearance, with the dark flight feathers contrasting with pale wing linings. Under ideal viewing conditions, two bare patches of blue skin at the base of the neck are diagnostic. *Habitat.* Wide ranging, but breeds on cliffs. Endemic to southern Africa and becoming rare. The most common vulture in the Drakensberg and Transkei. *Call.* Cackling and hissing noises given when feeding on carcasses, and at the nest. *Afrikaans.* Kransaasvoël.

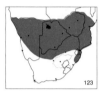

123 Whitebacked Vulture *Gyps africanus* L 95 cm
If seen from above or when banking steeply in flight, ad. has white lower back which contrasts with dark upperwings. This contrast is not as marked in Cape Vulture. Smaller than Cape Vulture and more of a bushveld species, avoiding mountainous cliffs. Imm. not distinguishable from imm. Cape Vulture unless smaller size can be judged. *Habitat.* Open savanna parkland. The most frequently seen vulture in bushveld game reserves, being common in the north. *Call.* Harsh cackles and hisses when feeding and at the nest. *Afrikaans.* Witrugaasvoël.

124 Lappetfaced Vulture *Torgos tracheliotus* L 100 cm
At close range the bare red skin on face and throat is diagnostic. In flight, the white thighs and white bar running along the forepart of the underwing from the carpal joint to the body, are conspicuous. Imm. is dark brown all over and most closely resembles the Hooded Vulture but is almost twice the size. *Habitat.* Thornveld with a preference for drier regions and is the vulture common in the Namib desert. Rare outside major game reserves. *Call.* Unknown. High-pitched whistles recorded during display. *Afrikaans.* Swartaasvoël.

125 Whiteheaded Vulture *Trigonoceps occipitalis* L 80 cm
In the region, the only dark vulture with large white wing patches. These patches are confined to the secondaries and are white in the female and off-white to grey in the male. The triangular-looking head and the neck are white, the naked face is pink, and the bill is orange with a blue base. Imm. is dark brown and distinguishable in flight from imm. Lappetfaced and Hooded vultures by the narrow white line between flight feathers and wing linings. *Habitat.* Thornveld and, more frequently, riverine forest in thornveld. Uncommon and rarely seen outside major game reserves. *Call.* Unknown. High-pitched chittering recorded when feeding. *Afrikaans.* Witkopaasvoël.

119 Bearded Vulture (Lammergeier) *Gypaetus barbatus* L 110 cm
Shape in flight is unlike any other vulture in the region: the long, narrow, pointed wings and long, wedge-shaped tail make the bird resemble a huge falcon. Ad. is mainly dark above with rust-coloured underparts and has a black mask across the face, terminating in a black 'beard'. (The 'beard' is not usually seen in the field.) Imm. is dark brown all over with underparts paling as it ages. *Habitat.* Remote high mountains; usually not found below 2 000 m. Confined to the Drakensberg's higher reaches. Thinly distributed and rare. *Call.* Silent except for high-pitched whistling during display. *Afrikaans.* Baardaasvoël.

19▲ 125▼ 123▼ 122▲

24▼

121 Hooded Vulture *Necrosyrtes monachus* L 70 cm

A small, nondescript brown vulture which could be confused with imm. Palmnut and Egyptian vultures. Distinguished from the former by its larger size, and down, not feather-covered head. Unlike this species, the imm. Egyptian Vulture has a long wedge-shaped, not square tail, bare facial skin and elongated nape feathers. *Habitat.* An uncommon thornveld species found mostly in game reserves in the north. *Call.* Normally silent. Emits soft whistling calls at nest. *Afrikaans.* Monnikaasvoël.

120 Egyptian Vulture *Neophron percnopterus* L 62 cm

Ad. could be confused with similar black and white Palmnut Vulture but this species has black primaries and a white, wedge-shaped tail. In flight, imm. looks like diminutive imm. Bearded Vulture, but has a bare face and a long, thin bill. Differs from imm. Palmnut and Hooded vultures by its long, wedge-shaped tail. *Habitat.* Could occur in almost any habitat throughout the region but is most usually found at kills in game reserves. Formerly common, it is now a vagrant with most recent sightings from the Transkei interior and Namibia. *Call.* Soft grunts and hisses when excited. *Afrikaans.* Egiptiese Aasvoël.

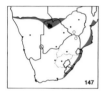

147 Palmnut Vulture *Gypohierax angolensis* L 60 cm

This small black and white vulture could be confused only with the Egyptian Vulture, from which it differs by having white, not black primaries, and by its tail being black and straight ended, not white and wedge shaped. Imm. resembles the Hooded Vulture but is smaller and has a feathered, not bare throat and neck. *Habitat.* Coastal forests in the vicinity of raffia palms. Extremely rare; restricted to Zululand and further north. *Call.* In flight sometimes utters a 'kok-kok-kok'. *Afrikaans.* Witaasvoël.

148 African Fish Eagle *Haliaeetus vocifer* L 63-73 cm

Apart from its haunting call, the ad. African Fish Eagle is unmistakable with its white head and breast, chestnut belly and forewings, black wings and white tail. Imm. identifiable by general dark brown coloration with white blotches on head and throat, and diagnostic white tail with dark terminal band. Overall impression in flight is of a large, broad-winged eagle with a short tail. *Habitat.* Aquatic: large rivers, lakes and dams. Coastal in some regions where it frequents estuaries and lagoons. Found throughout except in the extreme dry west. *Call.* A gull-like yelping given in flight or at rest. *Afrikaans.* Visarend.

170 Osprey *Pandion haliaetus* L 55-69 cm

Flight outline resembles that of a huge gull with white underwings, black patches at carpal joints and diagnostic black mask on white head. Might be confused with imm. African Fish Eagle found in similar habitat but this is a larger bird with broad wings and a very short tail. *Habitat.* Inland freshwater bodies, estuaries and lagoons, rarely fishing over open sea. Uncommon summer visitor. Found throughout the south and east but absent from the west, except for vagrants. *Call.* Silent in Africa. *Afrikaans.* Visvalk.

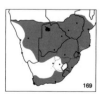

169 Gymnogene *Polyboroides typus* L 60-66 cm

Confusion could arise with the superficially similar Chanting Goshawks, but the broad floppy wings and single central white tail band should prove distinctive. Close views reveal bare yellow facial skin and elongated nape feathers. Imm. has no diagnostic plumage markings but the small head and broad wings, combined with the lazy manner of flight, help identification. At close range the imm. shows bare blackish facial skin. *Habitat.* Forests, riverine forests and open broad-leafed woodland. Uncommon and thinly distributed throughout except in dry desert areas. *Call.* During the breeding season it utters a whistled 'suuu-eeee-ooo'. *Afrikaans.* Kaalwangvalk.

72

147▲

120▲

121▲

148▼ 170▼ 169▲

142 **Brown Snake Eagle** *Circaetus cinereus* L 70-76 cm
Ad. distinguished from all other dark brown eagles by having unfeathered yellow legs. At rest, it appears unusually large headed with big yellow eyes. In flight the dark brown plumage contrasts with white primaries and secondaries. Imm. resembles ad. and differs from imm. Blackbreasted Snake Eagle by being darker brown, not rufous, and by the lack of barring on flight feathers. *Habitat.* Thornveld, avoiding open grasslands and forests. Rare in the south and thinly distributed in the east and north. *Call.* During the breeding season, a croaking 'hok-hok-hok-hok' is uttered in flight. *Afrikaans.* Bruinslangarend.

143 **Blackbreasted Snake Eagle** *Circaetus gallicus* L 63-68 cm
In flight, ad. resembles most closely the ad. Martial Eagle but differs by having white underwings, with primaries and secondaries barred black. At rest the large, dark-coloured head shows exceptionally large, bright yellow eyes. Imm. differs from ad. and the imm. Brown Snake Eagle by being a rich rufous colour, having barred flight feathers and an almost uniform dark brown tail. *Habitat.* Frequents a wide range of habitats, from true desert regions to open thornveld. Thinly distributed throughout except in the extreme south. *Call.* A melodious 'kwo-kwo-kwo-kweeu' *Afrikaans.* Swartborsslangarend.

145 **Western Banded Snake Eagle** *Circaetus cinerascens* L 55-60 cm
Differs from similar Southern Banded Snake Eagle by being stockier, less heavily barred below, and by having a shorter tail with a diagnostic broad white band across the middle, visible both in flight and at rest. In flight the tail band and white, not barred wing linings, differentiate it from the Southern Banded Snake Eagle. Imm. not easily distinguishable from imm. Southern Banded Snake Eagle. *Habitat.* Woodland bordering large rivers. Uncommon along the northern border. *Call.* A high-pitched 'kok-kok-kok-kok-kok'. *Afrikaans.* Enkelbandslangarend.

144 **Southern Banded Snake Eagle** *Circaetus fasciolatus* L 55-60 cm
Ad. resembles ad. Cuckoo Hawk but is much larger, with a large, rounded head and a relatively short tail. In flight distinguished from Western Banded Snake Eagle by its darker underwings and four, not two black bars on tail. *Habitat.* Coastal evergreen forests, riverine forests and open woodland in marshy areas. Uncommon on east coast as far south as St Lucia. *Call.* A harsh 'crok-crok-crok' and high-pitched 'ko-ko-ko-ko-keear'. *Afrikaans.* Dubbelbandslangarend.

146 **Bateleur** *Terathopius ecaudatus* L 55-70 cm
The most easily identified eagle of the region. The black, white and chestnut plumage combined with long wings and very short tail render this bird unmistakable. Flight action is diagnostic: with long wings held slightly angled, rarely flapping, it flies direct, canting from side to side. Imm. is a uniform brown and is distinguished from all other brown eagles by its flight action and very short tail. *Habitat.* Open thornveld and semi-desert. Avoids dense evergreen forests. Restricted to major game reserves in the east, and uncommon in the north and west. *Call.* A loud bark 'kow-wah'. *Afrikaans.* Berghaan.

139 **Longcrested Eagle** *Lophaetus occipitalis* L 52-58 cm
The combination of dull black plumage, white legs and long wispy crest render this eagle unmistakable. In flight shows conspicuous white bases to primaries and secondaries, and a black and white barred tail. Flight action is fast and direct on stiffly held wings with shallow wing beats. Imm. very like ad. but has short crest and grey, not yellow eyes. *Habitat.* Well-wooded country and forest edges, especially where marshy. Common in suitable habitat in the east and north-east. *Call.* High-pitched 'keee-ah' given during display flight or when perched. *Afrikaans.* Langkuifarend.

142 ▲   145 ▼ 144 ▼   143 ▲

                               146 ▼ 139 ▼

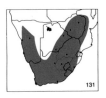

131 Black Eagle (Verreaux's Eagle) *Aquila verreauxii* L 75-95 cm
In flight, shows a diagnostic white 'V' on back, and a white rump. At rest the black plumage is relieved only by the two white lines down the back, and the yellow cere and feet. Imm. has diagnostic rust-coloured crown and nape which contrast with darker face and throat. The wing shape is characteristic: narrower at the base and broadening at the central secondaries. *Habitat.* Uncommon throughout the region. Prefers coastal cliffs and mountainous regions frequented by dassies. *Call.* Normally silent but does give a yelping 'keee-uup'. *Afrikaans.* Witkruisarend.

140 Martial Eagle *Polemaetus bellicosus* L 78-83 cm
The dark head, throat and upper breast combined with white, lightly spotted breast and belly, and very dark underwings are diagnostic in this huge eagle. From beneath, the ad. Blackbreasted Snake Eagle resembles this species but has a white underwing barred with black on primaries and secondaries. Imm. differs from imm. Crowned Eagle by having white, not spotted trousers and a more uniform underwing. *Habitat.* Uncommon throughout. Frequents a wide range of habitats, from desert to mountains, but is more common in thornveld. *Call.* Display call consists of a rapid 'klooee-klooee-klooee'. *Afrikaans.* Breëkoparend.

141 Crowned Eagle *Stephanoaetus coronatus* L 80-90 cm
The dark coloration, huge size and the shape identify this species. Ad. is dark grey above and rufous below, with breast and belly heavily mottled black. In flight shows well-rounded wings and a long tail. The underwing linings are rufous, with primaries and secondaries heavily barred black. Imm. similar to imm. Martial Eagle but differs in shape, has dark speckling on flanks and legs, and the tail and underwing are heavily barred. *Habitat.* Thinly distributed in dense evergreen forests and occasionally in open broad-leafed woodland in the east and north. *Call.* In display flights over forests a 'kewee-kewee-kewee' call is given. *Afrikaans.* Kroonarend.

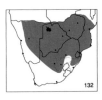

132 Tawny Eagle *Aquila rapax* L 65-80 cm
Much controversy surrounds the identification of this species and the migrant Steppe Eagle. Both species are very variable in colour, ranging from almost white to dark brown. At close range the gape length is diagnostic and in this species it extends only to below the eye. See wing shape in Steppe Eagle. *Habitat.* Thornveld and semi-desert areas. Common in the major game reserves but thinly distributed elsewhere in the east and north. *Call.* Silent except for grunts and croaks in aerial display. *Afrikaans.* Roofarend.

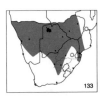

133 Steppe Eagle *Aquila nipalensis* L 65-80 cm
Plumage too variable for separation from Tawny Eagle. At close range and if clearly seen, the yellow gape is long and extends behind the eye, not to just below the eye as in Tawny Eagle. In flight, it shows a more curved 'S' trailing edge to the wing than the Tawny Eagle. *Habitat.* Thornveld and semi-desert regions. An uncommon summer visitor to the north. *Call.* Silent on its wintering grounds. *Afrikaans.* Steppe-arend.

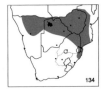

134 Lesser Spotted Eagle *Aquila pomarina* L 61-66 cm
A small dark eagle of buzzard-like proportions. Differs from Wahlberg's Eagle by having a short rounded tail. At rest the tightly feathered legs appear unusually thin and, seen at close range, the nostril is round, not oval. From above, imm. very similar to imm. Steppe Eagle, showing white bases to primaries, white edging to coverts and secondaries, a 'U'-shaped white rump and sometimes a white patch on back. *Habitat.* Often seen in company of Steppe Eagles over thornveld and frequently near water in open woodland. Uncommon summer visitor to the north and east. *Call.* Not recorded in this region. *Afrikaans.* Gevlekte Arend.

133▲

131▼

141▼

134▲

140▼

132▼

135 **Wahlberg's Eagle** *Aquila wahlbergi* L 55-60 cm
Usually a dark brown bird but has pale and intermediate phases. The flight shape is diagnostic with long, straight-edged wings and a long, narrow, square-ended tail. Pale phase birds could be confused with pale phase Booted Eagle but differ in flight shape and the long, pencil-thin tail. At rest this species shows a small pointed crest. *Habitat.* Frequents a wide range of habitats, from woodland to agricultural areas. Common summer visitor from central Africa to the north and east. *Call.* A loud 'keeee-ee' given in flight is most often heard. *Afrikaans.* Bruinarend (Wahlbergse Arend).

136 **Booted Eagle** *Hieraaetus pennatus* L 46-53 cm
This small, buzzard-sized eagle has two colour phases: during both of these it differs from Wahlberg's Eagle by its shorter, broader tail and broader wings. From above, a small white patch is visible at the base of the forewing, giving the impression of a pair of white 'braces'. The dark brown on head and face extends well below and contrasts with the small white throat.
Habitat. Thornveld and mountainous areas in the north and east. Dry mountains of the south and west, and desert regions of the west and north. Uncommon and thinly distributed throughout with a small number breeding in the south and west. *Call.* High-pitched 'kee-keeee' or 'pee-pee-pee-pee'. *Afrikaans.* Dwergarend.

138 **Ayres' Eagle** *Hieraaetus ayresii* L 45-55 cm
The bold black spotting on the white underparts, extending on to the belly and legs, help differentiate this species from the African Hawk Eagle. The underwing is much darker and lacks the marked dark central stripe of African Hawk Eagle. Diagnostic character in flight when seen from above is that Ayres' Eagle lacks the pale primary bases of African Hawk Eagle and has dark, uniformly coloured upperwings. Imm. unlikely to be distinguished from imm. African Hawk Eagle. *Habitat.* Well-wooded hilly country but avoids dense evergreen forests and arid areas. Occurs in the north and east but is rare, with most sightings during summer. *Call.* Normally silent but when displaying it utters a shrill 'pueep-pip-pip-pueep'. *Afrikaans.* Kleinjagerend.

137 **African Hawk Eagle (Bonelli's Eagle)** *Hieraaetus fasciatus* L 66-74 cm
Easily identified in flight by underwing pattern: a dark central stripe, which runs from the wing base to the carpal joint and along the forewing, contrasts with white leading and trailing edges. Differs from Ayres' Eagle by being less boldly spotted below and having pale bases to primaries. Imm. is russet below but still shows underwing character. *Habitat.* Open thornveld and woodland in hilly and rugged country. Avoids dense evergreen forests. Thinly distributed in suitable terrain in the centre and to the north. *Call.* During courtship a musical 'klee-klee-klee' is given. *Afrikaans.* Afrikaanse Jagarend.

158 **Black Sparrowhawk** *Accipiter melanoleucus* L 46-58 cm
The largest accipiter of the region. The black and white plumage, and the large size render this species unmistakable. A rare black form occurs but it has a white throat and this, combined with size, should preclude confusion with black Ovambo Sparrowhawk and Gabar Goshawk. Imm. has both pale and rufous phases. Pale phase may be confused with imm. African Goshawk but differs by lacking white eyebrow and black stripe down centre of throat. Rufous phase resembles imm. African Hawk Eagle but has more heavily streaked underparts and has a different shape and flight action. *Habitat.* Frequents a wide range of wooded areas; has adapted well to exotic plantations. Populations are increasing in the Transvaal. Found throughout the eastern half of the region in suitable habitat. *Call.* Normally silent except when breeding. Male's call is a 'keeyp', female's call is a short 'kek'. *Afrikaans.* Swartsperwer.

78

135 ▲   158 ▼   136 ▲

137 ▼   138 ▼

149 Steppe Buzzard *Buteo buteo* L 45-50 cm
Plumage coloration varies from pale brown through to almost black. It is generally darker below than the Forest Buzzard, and occurs mostly in different habitats. Frequently seen perched on telephone poles along roads and freeways. *Habitat.* A common and widespread summer visitor to open country, avoiding desert and well-wooded regions. *Call.* A gull-like 'pee-ooo'. *Afrikaans.* Bruinjakkalsvoël.

150 Forest Buzzard *Buteo tachardus* L 45-50 cm
The plumage is variable but less so than that of the similar, migrant Steppe Buzzard. The amount of white on the breast and belly varies, with some individuals showing a totally white throat, breast and belly. A common plumage is a white band across the breast distinctly demarcating the dark brown underparts. Imm. Steppe Buzzard often shows this character but it is less well defined. *Habitat.* Evergreen forests and forest edges. Has adapted to exotic eucalyptus and pine plantations. Uncommon in the east and in central mountain regions. *Call.* A gull-like 'pee-ooo'. *Afrikaans.* Bergjakkalsvoël.

152 Jackal Buzzard *Buteo rufofuscus* L 45-53 cm
Ad. has very dark plumage with bright chestnut breast and barred black and white belly. In flight vaguely resembles ad. Bateleur but the longer, rust-coloured tail and more flapping flight action eliminate confusion. Imm. readily identifiable by rufous-coloured underparts, wing linings and pale, short tail. Jackal Buzzards which show white breasts can be distinguished from the Augur Buzzard by their dark wing linings. *Habitat.* Generally confined to mountain ranges and adjacent grasslands. Common in suitable habitat in the south and central regions. Absent from the north where it is replaced by the Augur Buzzard. *Call.* A loud, drawn-out 'weeaah-ka-ka-ka', much like the yelp of the Blackbacked Jackal. *Afrikaans.* Rooiborsjakkalsvoël.

153 Augur Buzzard *Buteo augur* L 45-53 cm
Similar in shape and overall structure to Jackal Buzzard but differs by having white throat, breast and belly, and white wing linings. Some Jackal Buzzards show white breasts but their dark wing linings should eliminate confusion with this species. *Habitat.* Common in mountain ranges and hilly country in Namibia and Zimbabwe. *Call.* A harsh 'kow-kow-kow-kow' is given during display. *Afrikaans.* Witborsjakkalsvoël.

130 Honey Buzzard *Pernis apivorus* L 56 cm
Like the Steppe Buzzard which it resembles, this species is very variable in colour, but it differs by having two broad black bars at the base of its tail, and is more slender in appearance with a smaller, more compact head. At close range the scaly feathering which covers the face is diagnostic. Imm. lacks diagnostic tail pattern of ad. but the general shape and small head should help distinguish it from Steppe Buzzard. *Habitat.* Found mostly in woodland. An annual but rare summer visitor to the north-east. *Call.* 'Meeuuw', a higher pitched call than that of the Steppe Buzzard. *Afrikaans.* Wespedief.

151 Longlegged Buzzard *Buteo rufinus* L 51-66 cm
As with other buzzards, plumage is very variable. However, this species is large and bulky, and hovers occasionally in flight. The combination of dark belly, pale head, and longer and broader wings, also distinguish this from Steppe and Forest buzzards. Other useful identification characters are the white unmarked primaries with black wing tips and trailing edge, and the pale, almost translucent tail in the typical ad. Imm. inseparable from other imm. buzzards unless direct comparison of size and shape is made.
Habitat. Open grasslands and semi-desert. Rare summer vagrant, occurring in the north-west. *Call.* Typical buzzard 'pee-ooo'.
Afrikaans. Langbeenjakkalsvoël.

130▲   152▼   149▼   151▲

153▼   150▼

164 European Marsh Harrier *Circus aeruginosus* L 48-55 cm
Male easily identifiable by brownish body contrasting with grey on wings and tail. Female has creamy white cap and throat, and white-edged forewings. The imm. African Marsh Harrier, in worn and abraded plumage, can also show pale forehead and leading edge to wings but has a noticeably barred underwing and tail. Imm. similar to female or can have totally brown head. *Habitat.* Confined to marshy areas and adjoining fields. Rare vagrant in the north, with scattered records further south in central regions. *Call.* A plover-like 'clee-aa'. *Afrikaans.* Europese Paddavreter (E. Vleivalk).

165 African Marsh Harrier *Circus ranivorus* L 45-50 cm
Differs from female European Marsh Harrier by lacking the well-demarcated white crown and throat, and by having barring on flight feathers and tail. Distinguished from female Pallid and Montagu's harriers by its larger size, broader wings and by the absence of the small white rump.
Habitat. Marshland and flooded grassland and their adjacent fields. Common throughout the region except for the dry central and western districts.
Call. Normally silent except during breeding season, when various calls are given: 'woop' and 'chuk-chuk' being heard most often. *Afrikaans.* Afrikaanse Paddavreter (A. Vleivalk).

167 Pallid Harrier *Circus macrourus* L 38-46 cm
Smaller, and daintier in flight than Montagu's Harrier. Male easily recognized by very pale grey body and wings with black wing tips. Distinguished from male Montagu's Harrier by absence of streaking on belly and flanks and lack of black bar on secondaries. Female and imm. Pallid and Montagu's harriers are virtually indistinguishable in the field unless a very close view is obtained. The female and imm. of the Pallid Harrier, however, show whitish cheeks with a dark line through the eye and a black crescent behind the ear, a pattern lacking in the female and imm. Montagu's Harrier. *Habitat.* Upland grasslands and semi-desert. Uncommon summer visitor from northern hemisphere but more regularly encountered than Montagu's Harrier. Found throughout the region except in the south and west. *Call.* Silent in Africa. *Afrikaans.* Witborspaddavreter (Witborsvleivalk).

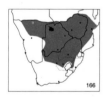

166 Montagu's Harrier *Circus pygargus* L 41-46 cm
Male differs from male Pallid Harrier by larger size and bulkier shape, streaks on belly and flanks, and a conspicuous black bar on secondaries. Female and imm. virtually inseparable from female and imm. Pallid Harriers: these are all usually referred to as 'Ringtail Harriers' because of the diagnostic barring on their tails. *Habitat.* Upland grasslands and over thornveld. Uncommon summer visitor from northern hemisphere. Distribution tends to the east and north of the region. *Call.* Silent in Africa. *Afrikaans.* Bloupaddavreter (Blouvleivalk).

168 Black Harrier *Circus maurus* L 48-53 cm
The pied plumage of the ad. male and female render this harrier unmistakable. Imm. resembles the imm. and female of the Pallid and Montagu's harriers, but differs by having white undersides to secondaries, primaries barred with brown, and a barred white and brown tail. *Habitat.* Open grasslands, scrub, and semi-desert and mountainous regions. Uncommon endemic to southern Cape, but outside the breeding season it wanders to northern and central regions. *Call.* A 'pee-pee-pee-pee' call is given during display and a harsh 'chak-chak-chak' when alarmed. *Afrikaans.* Witkruispaddavreter (Witkruisvleivalk).

82

165▲   168▼ 167▼

164▼ 166▼

162 Pale Chanting Goshawk *Melierax canorus* L 54-63 cm
Confusion only arises with Dark Chanting Goshawk where their ranges overlap. This species is a much paler grey, especially on the forewings, and has a pure white rump and white secondaries. Imm. is paler than imm. Dark Chanting Goshawk and has white, not barred rump. Superficially resembles a female harrier, but broader wings, flight action and long legs should eliminate confusion. *Habitat.* The common roadside hawk of the west and north-west, occurring in the dry west and desert. *Call.* Normally silent but, when breeding, it pipes a loud 'kleeuu-kleeuu-klu-klu-klu'. *Afrikaans.* Bleeksingvalk.

163 Dark Chanting Goshawk *Melierax metabates* L 50-56 cm
Much darker grey than Pale Chanting Goshawk. At rest the grey forewing does not contrast with the rest of the wing as in Pale Chanting Goshawk. In flight the white rump is noticeably barred: this is diagnostic, as are the grey, not white secondaries. Imm. has a darker brown breast band than imm. Pale Chanting Goshawk and has barring on the white rump. *Habitat.* Thornveld and open broad-leafed woodland. Thinly distributed in the extreme north and east of the region. *Call.* Piping call similar to that of Pale Chanting Goshawk, 'kleeu-kleeu-klu-klu'. *Afrikaans.* Donkersingvalk.

154 Lizard Buzzard *Kaupifalco monogrammicus* L 35-37 cm
Larger and bulkier than the similar Gabar Goshawk, this bird has a white throat with a black line down the centre, a white rump and a broad white tail stripe. Imm. similar to ad. but has buff edges to mantle feathers. *Habitat.* Uncommon resident in open broad-leafed woodland and thornveld in the east and north. *Call.* A whistled 'peoo-peoo' and a trilling 'klioo-klu-klu-klu-klu'. *Afrikaans.* Akkedisvalk.

160 African Goshawk *Accipiter tachiro* L 36-39 cm
Ad. male might be confused with Cuckoo Hawk but has short, rounded wings and long yellow legs, whereas, when in flight, the Cuckoo Hawk appears more falcon shaped. General colour resembles closely the Little Sparrowhawk, but its greater size and lack of white tail spots distinguish this species. Imm. differs from all other accipiters by its white throat with central black line.
Habitat. Common in thick evergreen forests, riverine forests in dry areas, and exotic plantations, in the south, east and north. *Call.* A short 'whit' or 'chip' is uttered in soaring flight over territory. Female has mewing 'kee-uuu' call. *Afrikaans.* Afrikaanse Sperwer.

128 Cuckoo Hawk *Aviceda cuculoides* L 40 cm
At rest resembles male African Goshawk but differs by having a crest, a grey throat and upper breast which end abruptly in a bib, and a heavily rust-barred breast and belly. In flight has long pointed wings, not rounded as in African Goshawk, with rufous wing linings. Imm. difficult to distinguish from imm. African Goshawk but has shorter legs, a small crest and in flight, pointed, not rounded wings. *Habitat.* A species which frequents well-wooded areas and thick woodland. Uncommon and thinly distributed in the east and north. *Call.* A loud, far-carrying 'teee-oooo' whistle and a shorter 'tittit-eooo'. *Afrikaans.* Koekoekvalk.

84

162▲

154▼

163▲

160▲

128▼

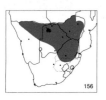

156 Ovambo Sparrowhawk *Accipiter ovampensis* L 33-40 cm
Best distinguished from similar Little Banded Goshawk by heavier barring below, grey-, not rufous-barred black and white tail, and orange, not yellow cere and legs. Black form differs from the black form of Gabar Goshawk by orange, not red cere and legs. The rufous imm. Ovambo Sparrowhawk differs from the ad. Redbreasted Sparrowhawk by its rufous, not dark grey head, and by the dark spot behind its eye. Pale imm. has a very pale head with dark ear coverts. *Habitat.* Open broad-leafed woodlands, avoiding thick forests and desert areas, and has adapted to exotic plantations, favouring poplars. Thinly distributed in the east and north. *Call.* A a soft 'keeep-keeep-keeep' is given when breeding. *Afrikaans.* Ovambosperwer.

155 Redbreasted Sparrowhawk *Accipiter rufiventris* L 33-40 cm
Ad. identified by uniformly rufous underparts, slate grey upperparts, lack of white rump and little barring on wings and tail. Imm. plumage very variable but is generally much darker above and more boldly marked below than imm. Ovambo Sparrowhawk, which lacks the uniform dark cap, showing only a dark spot behind the eye. *Habitat.* An inconspicuous species which has adapted to eucalyptus and pine plantations, especially in hilly areas or mountain ranges. Found throughout the southern, eastern and central regions in suitable plantations, and even in small stands in flat open areas. *Call.* A sharp, staccato 'kee-kee-kee-kee-kee' during display. *Afrikaans.* Rooiborssperwer.

161 Gabar Goshawk *Micronisus gabar* L 30-34 cm
Distinguished by its grey throat and breast, red cere and legs. Could be confused with Lizard Buzzard but that has a white throat with a central stripe and a broad white tail band. Black form Gabar differs from black form Ovambo Sparrowhawk by the red, not orange cere and legs. Imm. differs from other imm. accipiters by its white rump. *Habitat.* Frequents thornveld and open broad-leafed woodland, avoiding thick forest and desert areas. Uncommon throughout, but absent from the extreme south and west. *Call.* A high-pitched 'kik-kik-kik-kik-kik'. *Afrikaans.* Kleinsingvalk.

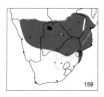

159 Little Banded Goshawk (Shikra) *Accipiter badius* L 28-30 cm
Stockier than the smaller Little Sparrowhawk and lacks the white rump and tail spots. Differs from Ovambo Sparrowhawk by its russet, not grey barring below, and yellow, not orange legs. Imm. differs from imm. Little Sparrowhawk by having mottled and barred rust-coloured underparts (not white underparts streaked with black), and by lacking the white tail spots. *Habitat.* Avoids dense evergreen forests and dry regions. Common in the east and north. *Call.* Male's call is a high-pitched 'keewik-keewik-keewik'. Female's call is a softer, mewing 'kee-uuu'. *Afrikaans.* Gebande Sperwer.

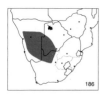

186 Pygmy Falcon *Polihierax semitorquatus* L 18-20 cm
Its small size is distinctive. Shrike-like, it sits very upright on exposed perches and hawks insects. Male is grey above and white below. Female has chestnut back. *Habitat.* Dry thornveld and semi-desert regions. Uncommon in the west and north-west, often found near Sociable Weaver nests. *Call.* A 'chip-chip' and other 'kik-kik-kik-kik' calls. *Afrikaans.* Dwergvalk.

157 Little Sparrowhawk *Accipiter minullus* L 23-25 cm
Distinguished from the similar Little Banded Goshawk and the much larger African Goshawk by its white rump and two white spots on the uppertail. Imm. very similar to imm. African Goshawk in colour but is much smaller, shows the white tail spots, and lacks the black stripe down the centre of its white throat. *Habitat.* Prefers open woodland, and has adapted to exotic plantations. Found in the east and north but absent from the drier south and west. *Call.* A high-pitched 'kek-kek-kek-kek' is uttered by the male. Female has a softer 'kew-kew-kew'. *Afrikaans.* Kleinsperwer. .

156▲   161▼   159▼   155▲

186▼   157▼

171 Peregrine Falcon *Falco peregrinus* L 34-38 cm
Confusable only with Lanner Falcon, from which it differs by having a smaller, more compact body, a black forehead and crown, a well-defined black moustachial stripe, and white underparts finely barred black. Its flight action is swifter and more agile than that of Lanner Falcon, and its more pointed wings and relatively shorter tail can be seen. Imm. is darker in appearance than imm. Lanner Falcon and has forehead and crown dark brown, not rufous.
Habitat. High cliffs and gorges, both coastal and inland. Although found throughout the region, resident population is very rare. The northern hemisphere race *calidus* is found in small numbers during summer.
Call. A loud, fast 'kek-kek-kek-kek-kek' is uttered over breeding territory.
Afrikaans. Swerfvalk.

172 Lanner Falcon *Falco biarmicus* L 40-45 cm
Rufous forehead and crown, thin moustachial stripe and pinkish unmarked underparts, distinguish this species from the Peregrine. It also has broader, more rounded wings, a longer tail and a floppier, less dynamic flight action than the Peregrine. Imm. much paler than imm. Peregrine with less streaking on underparts and sometimes with an almost white head. *Habitat.* A wide range of habitats, from mountainous terrain to deserts and open grassland. Usually avoids thick forests. Common throughout the region. *Call.* A harsh 'kek-kek-kek-kek-kek', not unlike a Peregrine's call. *Afrikaans.* Edelvalk.

129 Bat Hawk *Macheiramphus alcinus* L 45 cm
A dark brown bird, appearing black in the field, with a varying amount of white on throat and abdomen. Very falcon-shaped in flight and could easily be mistaken for a Lanner or Peregrine Falcon if seen in twilight when colours are not discernible. Wings are broader at the base and tail appears shorter than in either of those species and, when seen effortlessly catching bats in flight, this bird leaves no doubt as to its identification. *Habitat.* Woodland but not thick forests. Crepuscular and nocturnal, hiding in thick foliage during the day. Probably more common than presently thought as it is easily overlooked. Confined to the east and north. *Call.* A high-pitched, kestrel-like 'kek-kek-kek-kek-kek'. *Afrikaans.* Vlermuisvalk.

126 Black Kite (Yellowbilled Kite) *Milvus migrans* L 56 cm
Differs from similarly sized Steppe Buzzard by longer, forked tail and narrower wings which are more acutely angled backwards in flight. Although much the same size, the African Marsh Harrier has a longer, square-ended tail and flies with wings canted up. Black Kite has floppy flight action and twists its long tail slightly to the vertical for steering. The ad. in the African race *parasitus* has a yellow, not black bill. *Habitat.* Occurs in all habitats throughout the region. Regularly found in built-up areas. Common summer visitor from the Palearctic and central Africa. *Call.* A frequently emitted high-pitched squealing and whinnying. *Afrikaans.* Swartwou.

127 Blackshouldered Kite *Elanus caeruleus* L 33 cm
A small, easily identified grey and white raptor with diagnostic black shoulder patches. Has a characteristic habit of hovering and, when perching, it often flicks its short white tail. Imm. is browner than ad. with barred buff back and dark shoulders. *Habitat.* Found in a wide range of habitats from mountainous regions to agricultural lands and open thornveld, but avoids thick, evergreen forests. Common throughout the region. *Call.* Various whistles and harsher 'kek-kek-kek' sounds. *Afrikaans.* Blouvalk.

126▲   127▼ 171▼

129▼   172▼

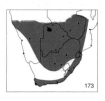

173 European Hobby *Falco subbuteo* L 28-35 cm
In flight the long pointed wings and relatively short tail give this species a swift-like appearance. The black head and moustachial stripes contrast boldly with the white throat, thus separating it from female and imm. Eastern Redfooted Falcon. Distinguished from similarly sized and shaped African Hobby by the lack of reddish brown on face and underparts. *Habitat.* Open broad-leafed woodland and thornveld. Avoids thick forests and deserts. Uncommon summer visitor throughout the region except for the extreme west and south-west. *Call.* Normally silent in Africa. *Afrikaans.* Europese Boomvalk.

174 African Hobby *Falco cuvierii* L 28-30 cm
Similar in size and shape to European Hobby but lacks the contrasting head markings and has a rust-coloured face, chest and underparts. Might be mistaken for Taita Falcon which is more robust and has rust nape markings and a white throat. *Habitat.* Open broad-leafed woodland, forests and adjoining open country. Rare in the north and north-east with vagrant records south to Natal and the eastern Cape. *Call.* A high-pitched 'kik-kik-kik-kik' is given during display. *Afrikaans.* Afrikaanse Boomvalk.

176 Taita Falcon *Falco fasciinucha* L 28-30 cm
This small, robustly built falcon is shaped like a Peregrine and, in flight, shows the dash and speed of that species. In coloration it resembles an African Hobby but differs by having rust-coloured patches on the nape, and a white throat which contrasts with black moustachial stripes. Imm. similar to ad. but has buff edges to back feathers. *Habitat.* High cliffs and gorges. Very rare, confined to gorges and mountains along the Zambezi River in Zimbabwe, and in Moçambique, and the highlands of these countries. *Call.* A high-pitched 'kree-kree-kree' and 'kek-kek-kek'. *Afrikaans.* Teitavalk.

177 Eleonora's Falcon *Falco eleonorae* L 38 cm
Almost as large as the Peregrine from whose pale phase this species can be distinguished by its rufous, heavily streaked underparts, and longer wings and tail. Differentiated from European Hobby by larger size, rufous underparts and lack of red 'trousers'. Dark phase most likely to be confused with rare black phase Sooty Falcon but is considerably larger and has greenish, not yellow legs and feet. Imm. differs from imm. Peregrine by having a much darker underwing. *Habitat.* Open broad-leafed woodland and adjoining grasslands. Rare vagrant during late summer. Only two records, both in the east. *Call.* Silent in our region. *Afrikaans.* Eleonoravalk.

185 Dickinson's Kestrel *Falco dickinsoni* L 28-30 cm
The combination of a grey body contrasting with a very pale grey head and white rump should rule out confusion with any other grey falcon in the region. Imm. very similar to ad. but has white barring on flanks. *Habitat.* Open palm savanna and a variety of wooded areas, very often in the vicinity of baobab trees. Uncommon and thinly distributed in the east. *Call.* High-pitched 'keee-keee-keee'. *Afrikaans.* Dickinsonse Valk (Grysvalk).

178 Rednecked Falcon *Falco chicquera* L 30-36 cm
This small falcon is unmistakable with its chestnut head and nape, and dark brown moustachial stripes. Upperparts are blue-grey, finely barred black, and the grey tail is tipped with a broad black band. Underparts are white, finely barred black, with a rufous wash across the breast. Imm. has a dark brown head, two buff patches on nape, and pale rust-coloured underparts finely streaked with brown. *Habitat.* Palm savanna but also found in drier areas in the west, and in tree-lined rivercourses in desert regions. Uncommon and thinly distributed in the western, central and north-eastern regions. *Call.* A shrill 'ki-ki-ki-ki-ki' during breeding season. *Afrikaans.* Rooinekvalk.

173▲   176▼   177▼   174▲

185▼   178▼

180 Eastern Redfooted Falcon *Falco amurensis* L 30 cm
Ad. male's dark slate plumage contrasts with the diagnostic white underwing linings. Female and imm. resemble European Hobby but have a white forehead and pale crown. Female lacks the rufous crown, nape and underparts of female Western Redfooted. *Habitat.* Open grasslands. Roosts communally in tall trees in towns. Common summer visitor to the east and north. *Call.* Silent in Africa. *Afrikaans.* Oostelike Rooipootvalk.

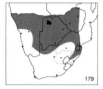

179 Western Redfooted Falcon *Falco vespertinus* L 30 cm
Ad. male is sooty grey with a chestnut vent and lacks the white underwing linings of Eastern Redfooted. Female differs from female Eastern Redfooted by having rufous crown, nape and underparts. Imm. might be confused with European Hobby but differs by having paler underparts and a much paler underwing with a dark trailing edge. *Habitat.* Open grasslands and thornveld, usually in dry areas. Common summer visitor to the north-west, with scattered records further south and east. *Call.* Normally silent in Africa. *Afrikaans.* Westelike Rooipootvalk.

175 Sooty Falcon *Falco concolor* L 32-35 cm
Differs from ad. male Western Redfooted Falcon by its grey vent and yellow, not red legs and cere. Flight shape like European Hobby but wings are longer. Imm. has shape of ad. but is pale grey below and heavily streaked. *Habitat.* Well-wooded coastal areas but also desert regions. From December to May, rare visitor on east coast, south to Natal and the eastern Cape. Vagrant to northern Cape and Namibia. *Call.* Silent in Africa. *Afrikaans.* Roetvalk.

184 Grey Kestrel *Falco ardosiaceus* L 30-33 cm
Distinguished from Sooty Falcon by shorter, broader wings, a larger, more obvious bill and stockier, more robust build. At rest the wings do not reach the tip of the square tail. An entirely grey falcon, it lacks the pale head and rump of Dickinson's Kestrel. *Habitat.* Open palm savanna. Vagrant to the extreme north and west. *Call.* Silent in our area. *Afrikaans.* Donkergrysvalk.

181 Rock Kestrel (Common Kestrel) *Falco tinnunculus* L 33-39 cm
Differs from Lesser Kestrel by having a spotted chestnut back and wings, lacking grey on secondaries and having the underwing spotted and barred, not silvery white. Female and imm. differ from female and imm. Lesser Kestrel, as that species is slimmer, is paler below, especially on underwing, and the tail tip, when closed, is slightly wedged. Differs from Greater Kestrel by smaller size, spotted back, and more heavily marked underwing. *Habitat.* A wide range, but usually mountainous or rocky terrain. Common throughout. *Call.* High-pitched 'kee-kee-kee' or 'kik-kik-kik'. *Afrikaans.* Kransvalk.

182 Greater Kestrel *Falco rupicoloides* L 36-40 cm
At close range the diagnostic whitish eye is obvious. Sexes similar and are distinguished from female Rock and Lesser kestrels by a grey, barred tail, whitish underwing, and a paler head which lacks moustachial stripes. Imm. is more rufous than imm. Rock and Lesser kestrels and has streaked, not spotted underparts. *Habitat.* Found throughout in drier regions, avoiding well-wooded areas. Absent from the south and the eastern lowlands. *Call.* During display a shrill 'kwirrr' or 'kweek' is given. *Afrikaans.* Grootrooivalk.

183 Lesser Kestrel *Falco naumanni* L 30 cm
Male differs from male Rock Kestrel by its unspotted chestnut back and obvious grey secondaries. Female and imm. very similar to female and imm. Rock Kestrel but are generally paler below, especially on the underwing, and have the tail tip slightly wedge shaped. *Habitat.* Open grasslands and agricultural areas. Roosts communally in tall trees in towns. Summer visitor absent from central west, and desert coasts. *Call.* Silent in Africa. *Afrikaans.* Kleinrooivalk.

92

179▲   175▼   180▲   183▼   184▼

181▼   182▼

195 Cape Francolin *Francolinus capensis* L 42 cm
Distinguished from all other large, dark francolins by having pale cheeks which contrast with a dark cap. Lacks any red on throat or around the eyes. Has obscure white streaking on lower belly. *Habitat.* The common francolin of fynbos. Has adapted to wheatfields but avoids exotic plantations. Common endemic to the extreme south and south-western Cape. *Call.* A loud, screeching 'cackalac-cackalac-cackalac'. *Afrikaans.* Kaapse Fisant.

194 Redbilled Francolin *Francolinus adspersus* L 35-38 cm
A dark francolin which has diagnostic bare yellow skin around the eyes. More uniform in colour than other dark francolins, is slightly paler below and lacks any streaking or blotching. *Habitat.* Dry thornveld and open broad-leafed woodland in the drier regions of the north and west. Feeds freely in the open and is less skulking than other francolins. *Call.* A loud, harsh 'chaa-chaa-chek-chek' is uttered at dawn and dusk. *Afrikaans.* Rooibekfisant.

196 Natal Francolin *Francolinus natalensis* L 35 cm
Similar to the larger Rednecked Francolin but does not have bare red skin around eyes and on throat. Generally brown above and speckled black and white below. Imm. similar to ad. but has less speckling below and is darker brown on breast and back. *Habitat.* Riverine scrub, stony outcrops covered in scrub, and adjoining grassland. Uncommon in the east and north-east. *Call.* A ringing 'kwali-kwali-kwali' given at dawn and dusk. *Afrikaans.* Natalse Fisant.

199 Swainson's Francolin *Francolinus swainsonii* L 38 cm
The only large brown francolin with black-brown legs. Lack of white markings on underparts, its black bill and dark legs distinguish it from Rednecked Francolin. Resembles closely the Cape Francolin but confusion is unlikely because their ranges never overlap. *Habitat.* Dry thornveld and agricultural lands. Common endemic to the central and northern regions. Seen in small groups of three to five. *Call.* A raucous 'krraae-krraae-krraae' given by the males at dawn and dusk. *Afrikaans.* Bosveldfisant.

198 Rednecked Francolin *Francolinus afer* L 36 cm
Similar to Swainson's Francolin in having bare red skin on throat and around eyes but has red, not brown legs and has conspicuous white striping on flanks. The six races are very variable in plumage coloration but the red legs consistently distinguish this species from Swainson's Francolin.
Habitat. Evergreen temperate forests, their edges and adjoining grasslands. Common in the east of the region. *Call.* A loud 'kwoor-kwoor-kwoor-kwaaa' given at dusk and dawn. *Afrikaans.* Rooikeelfisant.

94

194▲

199▼

195▲   196▼

198▼

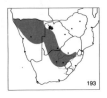

193 Orange River Francolin *Francolinus levaillantoides* L 35 cm
This species is slightly smaller than the very similar Redwing Francolin: it differs by having a thin, dark necklace which never broadens to form a dark breast band. There are many subspecies which vary greatly in coloration. Distinguished from Shelley's Francolin by lack of black barring on belly, and from Greywing Francolin by the white throat. *Habitat.* Stony ridges in dry grassland, and thornveld. Uncommon and thinly distributed in central and western regions. *Call.* A screeching 'weecheele-wecheele-weecheele'. *Afrikaans.* Kalaharipatrys.

192 Redwing Francolin *Francolinus levaillantii* L 38 cm
The black-speckled necklace on the breast is much broader than in any other francolin. It forms a distinct breast band, which distinguishes the Redwing from the Orange River Francolin. Imm. paler than ad. but still shows dark breast band. *Habitat.* Grasslands and meadows in mountainous terrain, usually on lower slopes and in valleys. Locally common in the south and east. *Call.* A piping 'pip-pip-peeep-peeep-peeep' is given, the last three notes being higher pitched. *Afrikaans.* Rooivlerkpatrys.

190 Greywing Francolin *Francolinus africanus* L 33 cm
A nondescript small grey francolin, easily identified by its grey freckled throat (not white or buff as in other small francolins). Sexes are similar and imm. is duller version of ad. but still displays throat character. *Habitat.* Mountainous regions in most of its range but coastal in the southern and western Cape. From the southern Cape the range extends to inland Natal and the highlands of Swaziland. Common but furtive and usually found in small coveys. *Call.* A high-pitched 'chee-chee-chleeoo' is given at dawn and dusk. *Afrikaans.* Bergpatrys.

191 Shelley's Francolin *Francolinus shelleyi* L 33 cm
The bright chestnut breast, chestnut-striped flanks, and black and white barred belly are diagnostic in male. Female superficially resembles smaller female Coqui Francolin but lacks the white eye-stripe and the breast is chestnut striped. Imm. resembles ad. female but has throat streaked brown. *Habitat.* Stony ridges and adjacent grasslands, and in lightly wooded areas. Locally common in suitable habitat in the east and north. *Call.* A 'klee-klee-kleer' phrase is repeated many times at dawn and dusk. *Afrikaans.* Laeveldpatrys.

193▲　　　　　　　　**192▼**　　　　　　　　**190▼**

191▼

187 Chukar Partridge *Alectoris chukar* L 33 cm
The creamy white throat with broad black border and barred chestnut, black and white flanks, are diagnostic. Shelley's Francolin has a white throat bordered by a freckled necklace but lacks the bright vertical barring on flanks. *Habitat.* Introduced into different parts of the region but now occurs only on Robben Island. *Call.* A distinctive 'chuk-chuk-chuk-chukar'. *Afrikaans.* Asiatiese Patrys.

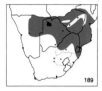

189 Crested Francolin *Francolinus sephaena* L 33-35 cm
The dark cap with contrasting broad white eyebrow stripe and dark-striped chest are diagnostic. Female is more barred above and imm. has buff, not white eyebrow stripe. The tail is frequently held cocked at a 45° angle. *Habitat.* On the ground under heavy, tangled growth and along dry rivercourses in thornveld. Common in the north and east. *Call.* A rattling 'chee-chakla, chee-chakla' given near roosts. *Afrikaans.* Bospatrys.

188 Coqui Francolin *Francolinus coqui* L 28 cm
The only francolin in the region to show a plain buffy head with a contrasting darker crown. Female resembles Shelley's Francolin but has a broad white eye-stripe which runs behind the eye and on to the neck. Much smaller than most francolins. *Habitat.* Grasslands in lightly wooded areas and thornveld. Uncommon and localized in suitable habitat in the north and east. *Call.* Easily located and identified by its six-part 'kwee-eek-eek-eek-eek-eek' call. *Afrikaans.* Swempie.

197 Hartlaub's Francolin *Francolinus hartlaubi* L 26 cm
The smallest francolin of the region. Male is easily identified by its dark cap contrasting with its white eyebrow, and pale underparts heavily streaked with brown. Female and imm. are dull brown birds lacking any distinguishing features. *Habitat.* Boulder-strewn slopes and rocky outcrops in hilly and mountainous regions. Endemic to central Namibia. Uncommon and local in suitable habitat. Often seen in small groups scurrying over rocks. *Call.* A 'wa-ak-ak-ak-ak' alarm call. *Afrikaans.* Klipfisant.

98

87▲ 197▼ 189▲ 188▼

200 **Common Quail** *Coturnix coturnix* L 18 cm
A small gamebird usually seen in flight, when its dumpy body and whirring wings attract attention. On the ground appears rodent-like, running swiftly through the grass in a hunched position. Male shows black throat but lacks the black breast band of male Harlequin Quail. Imm. and female are much paler below than female Harlequin Quail. *Habitat.* Grasslands and fields. Chiefly a summer visitor throughout, being rare or absent from extreme north-eastern and desert regions. *Call.* A 'whit-tit-tit' and a 'crwee-crwee' in flight. *Afrikaans.* Afrikaanse Kwartel.

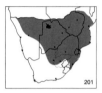

201 **Harlequin Quail** *Coturnix delegorguei* L 18 cm
Much darker in appearance than Common Quail. Male differs from Common and Blue quails by its chestnut underparts with black streaks and its black breast. Female and imm. are pale versions of male but are almost indistinguishable from Common Quail in flight. *Habitat.* Open grasslands, damp fields and grassland adjoining freshwater ponds. Summer visitor to the northern, eastern and central regions. More abundant in some years and nomadic, appearing commonly in various areas for short periods and then departing. *Call.* Easily located by the male's ringing 'vet-veet-veet'. *Afrikaans.* Bontkwartel.

202 **Blue Quail** *Coturnix adansonii* L 15 cm
Male unmistakable. It has a black and white face and throat pattern, combined with russet and dark blue underparts which, in flight, appear black. On the ground, female and imm. can be distinguished from other quails by their barred underparts. The smallest quail of the region, similar in size to the buttonquails. *Habitat.* Damp and flooded grasslands in lightly wooded areas. A rare summer visitor to the east. *Call.* Described as a high-pitched whistle 'teee-ti-ti'. *Afrikaans.* Bloukwartel.

205 **Kurrichane Buttonquail** *Turnix sylvatica* L 14 cm
In flight the lack of dark rump and back distinguish this species from Blackrumped Buttonquail. The wings and back appear pale and uniformly coloured. On the ground the black flecks on the breast and flanks, and the absence of chestnut on the face, are diagnostic. *Habitat.* Open grasslands, grasslands in lightly wooded areas, and agricultural lands. The most frequently encountered buttonquail throughout the region, except in the south and south-west. *Call.* A soft, two-note 'dooo-dooo' is thought to be given by female only. *Afrikaans.* Bosveldkwarteltjie.

206 **Blackrumped Buttonquail** *Turnix hottentotta* L 14 cm
This and the Kurrichane Buttonquail, unless clearly seen, are extremely difficult to distinguish. In flight, this species shows a diagnostic dark back and rump which contrast with paler wings. Both species are noticeably smaller than the similar *Coturnix* quails and appear to buzz through the air when put to flight. *Habitat.* Moist grasslands in hilly and mountainous areas. A rare and little-known species recorded from scattered localities in the south and east. *Call.* A flufftail-like 'ooooop-ooooop'. *Afrikaans.* Kaapse Kwarteltjie.

100

200 ▲

201 ▲

202 ▼

205 ▼

206 ▼

203 Helmeted Guineafowl *Numida meleagris* L 56 cm
The grey body, flecked with white, the naked blue and red head, and bare casque on the crown, render this large gamebird unmistakable. Imm. like ad. but has a less developed casque on crown, browner body coloration with white flecking enlarged on neck feathering. *Habitat.* Grasslands, broad-leafed woodlands, thornveld and agricultural lands. A common, conspicuous bird found throughout the region except in the extreme dry west. *Call.* A loud, repeated 'krrdii-krrdii-krrdii-krrdii' and a 'kek-kek-kek-kek' alarm note. *Afrikaans.* Gewone Tarentaal.

204 Crested Guineafowl *Guttera pucherani* L 50 cm
A black guineafowl finely spotted with white, and having a tuft of black feathers on the crown. The naked face is blue-grey with a large bare white patch on ears and nape. In flight shows a white patch on primaries. Imm. is mottled and barred above and below. *Habitat.* Thick evergreen forests, and broad-leafed woodland with dense secondary growth. Uncommon and local in the north and east. *Call.* A 'chik-chik-chil-chrrrrr' is given by males. A soft 'keet-keet-keet' contact call is uttered by groups foraging on the forest floor. *Afrikaans.* Kuifkoptarentaal.

228 Redknobbed Coot (Crested Coot) *Fulica cristata* L 44 cm
A black, duck-like bird with a white bill and a white unfeathered forehead. The two red knobs on the forehead are visible only at close range except during the breeding season, when they swell and become quite noticeable. Imm. is dull brown and differs from the smaller imm. Moorhen by lacking white undertail coverts. *Habitat.* Virtually any stretch of fresh water except fast-flowing rivers. Common throughout the region. *Call.* A harsh, metallic 'claak'. *Afrikaans.* Bleshoender.

229 African Finfoot *Podica senegalensis* L 63 cm
When swimming, the head shape and submerged body suggest a Darter, but this species differs by having a much shorter, stouter neck. Out of the water the bright orange feet and legs are conspicuous. *Habitat.* Densely vegetated rivers with overgrown banks. An uncommon, furtive species found on rivers in the east and north. *Call.* Normally silent except for a short, frog-like 'krork'. *Afrikaans.* Watertrapper.

240 African Jacana *Actophilornis africanus* L 28 cm
Unmistakable. A rufous bird with a white breast and yellow collar, and a contrasting black and white head which highlights the blue shield. The extremely long toes and nails which enable it to walk over floating vegetation are easily visible. Female is slightly larger than male, and imm. is a duller version of the ad. with a paler belly. Takes to flight readily, flying with its long legs dangling, and on alighting raises both wings above its back. *Habitat.* Any body of water containing floating vegetation but it also frequents flooded grasslands. While common in suitable habitat in the north and east, it occurs sporadically throughout the region. *Call.* A sharp, ringing 'krrrek' and a 'krrrrrrk' flight call. *Afrikaans.* Grootlangtoon.

241 Lesser Jacana *Microparra capensis* L 15 cm
A much smaller bird than African Jacana, it has a white belly, breast and throat but otherwise is pale brownish. Resembles an imm. African Jacana but the black on the crown and the blue shield are absent. Imm. is smaller and lacks the blackish nape of African Jacana. In flight the ad. shows white on the trailing edges of the secondaries which contrast with the dark primaries. *Habitat.* Quiet stretches of water containing aquatic plants amongst which it can forage. Occurs in the eastern and northern sectors of the region but it is a rare species. *Call.* A soft, often-repeated 'krick'. *Afrikaans.* Dwerglangtoon.

102

203 ▲ 228 ▼ 229 ▼ 204 ▲

240 ▼

241 ▼

223 Purple Gallinule *Porphyrio porphyrio* L 46 cm

The large size, massive red bill and frontal shield, long red legs and toes combined with its general purplish coloration and turquoise neck and breast, are unmistakable. The Lesser Gallinule is half the size and has a green, not red frontal shield. Imm. is dull brown but still shows the massive bill. *Habitat.* Thick stands of reedbeds and flooded grasslands. Common throughout the region except in dry western regions. *Call.* Variety of harsh shrieks, booming notes and a 'keyik' contact call. *Afrikaans.* Grootkoningriethaan.

224 Lesser Gallinule (Allen's Reedhen) *Porphyrula alleni* L 25 cm

Easily distinguished from Moorhen by smaller size, green, not red frontal shield, lack of white stripes along flanks and red, not green legs. Imm. differentiated from imm. Moorhen by pale, fleshy coloured, not greenish brown legs, and by lacking the white stripes along flanks. Very similar to imm. American Purple Gallinule but that species has olive-coloured legs. *Habitat.* Thick stands of reedbeds and flooded grasslands. Common in the northern part of the region, becoming rarer further south and east. *Call.* A metallic 'klaark' and other frog-like noises. *Afrikaans.* Kleinkoningriethaan.

225 American Purple Gallinule *Porphyrula martinica* L 33 cm

Similar to Purple Gallinule but is smaller, has bright yellowish green, not red legs and a yellow tip to its bill. Differs from imm. Moorhen by lack of white stripes on flanks and has olive, not green legs. Imm. Lesser Gallinule very similar but that species has flesh-coloured legs. *Habitat.* Usually thick reedbeds but virtually anywhere, some having been found walking along roads and beaches in an exhausted state. Rare vagrant during summer to the west coast, with one record from the Natal coast. *Call.* Silent in South Africa. *Afrikaans.* Amerikaanse Koningriethaan.

226 Moorhen *Gallinula chloropus* L 32 cm

Unlike the glossy purple and green Lesser and American Purple gallinules, the Moorhen is dull black, and also differs by having green legs and a red frontal shield. Imm. is distinguished from imm. Lesser Moorhen and all other imm. gallinules by the white stripes on its flanks. Bolder than Lesser Moorhen and the gallinules, it swims freely in patches of open water. *Habitat.* Virtually any stretch of fresh water surrounded by reeds and tall grasses. Common throughout the region. *Call.* A sharp 'krrik'. *Afrikaans.* Grootwaterhoender.

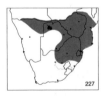

227 Lesser Moorhen *Gallinula angulata* L 24 cm

Very similar to the Moorhen but is noticeably smaller, has a yellow, not red bill, and lacks the white flank stripes. Imm. differs from imm. Moorhen and all other small gallinules by having a russet brown head and neck. *Habitat.* Thick reedbeds and flooded grasslands. More skulking than Moorhen, venturing infrequently into open water. Uncommon in the north and east. *Call.* Similar to Moorhen's 'krrrik', but higher pitched. *Afrikaans.* Kleinwaterhoender.

213 Black Crake *Amaurornis flavirostris* L 20 cm

Ad. is unmistakable with matt black coloration, bright yellow bill, red eyes and legs. Imm. is a grey version of ad. and has a black bill and dull red legs. A noisy bird, more often heard than seen, but it frequently ventures into the open, especially at dawn and dusk. *Habitat.* Marshes and swamps with a thick cover of reedbeds and other aquatic vegetation. Common in suitable habitat throughout the region except the dry west. *Call.* A throaty 'chrrooo' and rippling trill 'weet-eet-eet-eet'. *Afrikaans.* Swartriethaan.

223 ▲

226 ▼ 225 ▼

224 ▲

227 ▼ 213 ▼

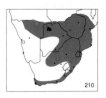

210 **African Rail** *Rallus caerulescens* L 37 cm
A distinctive species with a long, slightly decurved red bill and red legs, combined with a blue throat and breast, and chestnut upperparts. Imm. has a long bill and a buff throat and breast. More frequently seen than other rails and crakes, readily venturing out into the open, especially in early morning. *Habitat.* Marshes, thick reedbeds and flooded grasslands. Common throughout except in the dry central and western regions. *Call.* A high-pitched, trilling 'trrreee-tee-tee-tee-tee-tee'. *Afrikaans.* Grootriethaan.

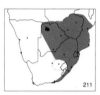

211 **Corncrake** *Crex crex* L 35 cm
Much paler sandy colour than the smaller African Crake. Although rarely seen, it is easily flushed and flies away on whirring wings, legs dangling, showing its diagnostic, conspicuous chestnut-orange wing coverts. *Habitat.* Open grasslands and lightly wooded, grassy areas. Summer visitor from Europe, arriving in November and departing in March. Uncommon in the north and east. *Call.* Silent when in Africa. *Afrikaans.* Kwartelkoning.

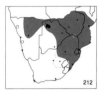

212 **African Crake** *Crex egregia* L 22 cm
Distinguished from the African Rail by its short stubby bill, and from the Spotted Crake by its more boldly barred flanks and belly. When flushed, the bird flies a short distance, legs dangling, showing clearly the brown mottled upperparts and black and white barred flanks. Imm. resembles ad. and, although browner in appearance, still shows black and white barred flanks. *Habitat.* Usually associated with rank vegetation surrounding fresh water, and open grasslands. Has adapted to sugar-cane plantations bordering rivers and dams. Uncommon in the north and east. *Call.* A high-pitched whistling has been described. *Afrikaans.* Afrikaanse Riethaan.

214 **Spotted Crake** *Porzana porzana* L 20 cm
Most likely to be confused with the African Crake but this is a much darker bird showing a yellow bill with a red base, and obvious white spots and stripes on upperparts. When flushed, the diagnostic greenish coloured legs and the barring on the flanks (less bold than on the African Crake) can be seen. When on the ground, the bird flicks its tail, thus making the buffy undertail coverts noticeable. *Habitat.* Flooded grasslands and thick reedbeds. A rare summer visitor from Europe, with scattered records in the north and east. *Call.* Silent when in Africa. *Afrikaans.* Gevlekte Riethaan.

216 **Striped Crake** *Aenigmatolimnas marginalis* L 20 cm
The rich brown upperparts with long white stripes on the back and wings are obvious characters. Undertail and flanks are russet. Female differs markedly from male by having a blue-grey breast and belly but still shows the rich russet flanks and undertail. *Habitat.* Flooded grasslands and marshes with short water grasses. Status not clear, with scattered records in the north and east. *Call.* A sharp 'tak-tak-tak' at night and on dark overcast days. *Afrikaans.* Gestreepte Riethaan.

215 **Baillon's Crake** *Porzana pusilla* L 18 cm
The smallest of the crakes, similar to a flufftail in size. Resembling a diminutive African Crake, it differs by having warm brown upperparts flecked and spotted with white. *Habitat.* Thick reedbeds, flooded grasslands and flooded broad-leafed woodland with long grass growth. Rarely seen in the open except at dawn or dusk. Uncommon throughout, and absent from the dry central and western regions. *Call.* A soft 'qurrr-qurrr' and various frog-like croaks. *Afrikaans.* Kleinriethaan.

210▲

212▼ 214▼

211◢

216▼ 215▼

217 Redchested Flufftail *Sarothrura rufa* L 16 cm
The only flufftail in the region to have the rufous head colour extend to the back and lower breast. In flight over reedbeds distinguished from Striped Flufftail by its black, not red tail. Female distinguished from female Striped Flufftail by appearing much blacker and larger, with a floppier, less deliberate flight. *Habitat.* Thick stands of reeds and tall water grasses. Common in marshes and even small vleis in the north and east. *Call.* A low, far-carrying 'ooouup-oooup' and a ringing 'klee-klee-klee-klee' given throughout the year but more readily during summer. *Afrikaans.* Rooiborsvleikuiken.

218 Buffspotted Flufftail *Sarothrura elegans* L 16 cm
The dark body is more speckled than other flufftails' and the chestnut extends only as far as the upper breast, not on to the lower breast as in Redchested Flufftail. If the male is seen on the forest floor, the impression is of a small, plump, chestnut-headed bird, heavily spotted with golden buff. Female and imm. are buff breasted and very dark brown above. *Habitat.* Thick evergreen forests and coastal scrub. Common in suitable habitat, ranging from Cape Town, north and east to Moçambique. *Call.* A low, foghorn-like 'dooooooo' given mainly at night and on overcast days. *Afrikaans.* Gevlekte Vleikuiken.

219 Streakybreasted Flufftail *Sarothrura boehmi* L 15 cm
The chestnut head and neck contrasting with the very pale throat are diagnostic characteristics. Most likely to be confused with Redchested Flufftail but, when flushed, appears much paler and it can be seen that the chestnut on the head does not extend on to the back and lower breast. Female is much paler below than female Redchested. *Habitat.* Swamps, reedbeds and flooded grasslands. Rare visitor to the extreme north during summer rainy season. *Call.* Repeated, double-note 'gawooo-gawoo'. *Afrikaans.* Streepborsvleikuiken.

220 Longtoed Flufftail *Sarothrura lugens* L 15 cm
This species is the darkest of the flufftails: the head and upper neck are deep chestnut, appearing almost black in the field. It differs from the similar Striped Flufftail by its black, not chestnut tail. Female differs from female Striped Flufftail by being much darker above and below. Imm. resembles female. *Habitat.* Upland grasslands and moist, grassy hollows. Rare and little-known species in our region, found only in the Eastern Highlands of Zimbabwe. *Call.* A crescendo call 'doh-doh-doh'. *Afrikaans.* Bronskopvleikuiken.

222 Whitewinged Flufftail *Sarothrura ayresi* L 14 cm
When flushed, both sexes show diagnostic square white panels on the secondaries. Flight is very fast and direct with whirring wing beats, ending with a sudden crash into a reedbed. On the ground, a thin white line is noticeable on the outer web of the first primary. Distinguished from Striped Flufftail by this white line, and by the barred black and chestnut tail. *Habitat.* Thick reedbeds in upland marshes and vleis, and shorter water grasses. Very rare with localized populations in Natal and the Transvaal. Scattered records from further north. *Call.* Not known. *Afrikaans.* Witvlerkvleikuiken.

221 Striped Flufftail *Sarothrura affinis* L 15 cm
Female differs from female Redchested Flufftail by being smaller, much swifter on the wing and having the tail suffused with chestnut. On the ground, the male's chestnut tail is diagnostic and distinguishes it from the male Whitewinged Flufftail, which has a barred chestnut and black tail. *Habitat.* Upland grasslands in mountainous regions and upland flooded meadows and marshes. Probably more common than presently thought, it is found in the south and east. *Call.* Similar to Buffspotted Flufftail in quality but shorter: 'dooo-dooo-dooo'. Also a ringing 'klinng-klinng-klinng'. *Afrikaans.* Gestreepte Vleikuiken.

217▲   218▼

222▲   220▼

219▲   221▼

1 Ostrich *Struthio camelus* H 2 m
Black and white male is the only bird which is as large as a man. In the female, the black plumage is replaced by brown. Chick resembles a korhaan but has a flattened bill and thick legs. Imm. resembles a small female. *Habitat.* Feral stock found on farms virtually throughout the region. The only genuine wild ostriches occur in northern Namibia and the Kalahari. All others are a mixture of races introduced from North Africa for domestic stock purposes. *Call.* Nocturnal, a booming leonine roar. *Afrikaans.* Volstruis.

463 Ground Hornbill *Bucorvus leadbeateri* L 90 cm
Unmistakable, turkey-sized black bird with conspicuous red face and throat patches. Walks on its tiptoes. In flight shows broad white wing patches. The female is distinguished from the male by having a blue throat patch. Imm. differs from ad. by having a yellow, not red face and throat patch. *Habitat.* Open grassy areas in thornveld and broad-leafed woodland, and in upland grassland. Common in certain areas, especially the larger reserves. Found in the north, east and south-east. *Call.* A loud, booming 'ooomph ooomph' early in the morning. *Afrikaans.* Bromvoël.

118 Secretarybird *Sagittarius serpentarius* L 140 cm
This bird's peculiar shape and long legs render it likely to be confused only with a crane at long range. In flight the two elongated central tail feathers project well beyond the main tail and the legs, producing an unmistakable flight shape. Imm. resembles ad. but has a shorter tail and yellow, not red bare facial skin. *Habitat.* Open grasslands from coastal regions to high altitudes. Avoids thick bush and forests. Found throughout the region except in arid zones. Now absent from many settled areas. *Call.* Normally silent but during display utters a deep croak. *Afrikaans.* Sekretarisvoël.

208 Blue Crane *Anthropoides paradisea* L 100 cm
The unusually large head, long slender neck and elongated, trailing inner secondaries are diagnostic. Sexes alike but female slightly smaller. Imm. lacks the long inner secondaries, is paler grey and has a faint chestnut wash on the head. *Habitat.* Usually associated with freshwater areas and open grasslands but has adapted to agricultural lands. Found in small groups or in pairs but sometimes non-breeding birds gather in large flocks. A common endemic found in the southern, central and north-western regions. *Call.* A loud, nasal 'kraaaank-kraaank' is given. *Afrikaans.* Bloukraanvoël.

209 Crowned Crane *Balearica regulorum* L 105 cm
Unlikely to be confused with any other species. In flight the large white upperwing patches are diagnostic. Imm. lacks the large, unfeathered, white face patch and the bristly crown is less well developed. During the non-breeding season, may be found in mixed flocks with Blue Cranes. *Habitat.* Marshes, dams and adjoining grasslands. Also in coastal lagoons in Transkei. Common but localized in the east and north. *Call.* A trumpeting flight call 'may hem' and a deep 'huum huum' when breeding. *Afrikaans.* Mahem.

207 Wattled Crane *Grus carunculatus* L 120 cm
This enormous crane with its long, white feathered wattles is unmistakable. At very long range can be easily identified by its dark face, and white head and neck. Imm. lacks white wattles. *Habitat.* Highland vleis and marshes and adjoining grasslands, as well as the larger water systems in the north. Usually found in pairs or small groups but non-breeding birds sometimes gather in flocks of 50 or more. Rare in the east and north. *Call.* A loud 'kwaarnk' is used as a contact call and in flight. *Afrikaans.* Lelkraanvoël.

209 ▲   207 ▲

208 ▲   1 ▼   463 ▼   118 ▼

230 **Kori Bustard** *Ardeotis kori* L 135 cm
The largest bustard of the region. The size and lack of any red on the nape and hind neck are diagnostic. Female is similar to male but is considerably smaller. Strides over the veld, swinging its head and neck with a peculiar backwards and forwards movement. Reluctant to fly except when threatened. *Habitat.* Dry thornveld, grassland and semi-desert. Range much restricted as a result of pressures of hunting and, consequently, it is not often found outside major game reserves. *Call.* A deep, resonant 'oom-oom-oom' is given during the breeding season. *Afrikaans.* Gompou.

231 **Stanley's Bustard (Denham's Bustard)** *Neotis denhami* L 104 cm
The dark cap, pale grey neck and chest, and conspicuous white wing markings are diagnostic. Female and imm. show less white in the wings and have brown speckling on the neck. The similar Ludwig's Bustard has a dark throat and breast, lacks the dark cap and has less white in the wing. *Habitat.* Open grassland and agricultural lands. Uncommon in the south and east. *Call.* A 'wak-wak' call and a booming 'oomp-oomp' are given. *Afrikaans.* Veldpou.

232 **Ludwig's Bustard** *Neotis ludwigii* L 90 cm
Most likely to be confused with the larger Stanley's Bustard but differs by having a dark brown neck and breast, lacking the dark cap and having far less white in the wings. Female noticeably smaller than male. In flight, appears much darker with very little white in the wings compared to Stanley's Bustard. *Habitat.* Much drier regions than Stanley's Bustard and found in the arid zones of the west. An uncommon and nomadic endemic to the south and west. *Call.* A short, hollow bark 'wuuk', repeated every six seconds. *Afrikaans.* Ludwigse Pou.

235 **Karoo Korhaan** *Eupodotis vigorsii* L 58 cm
Plumage very variable but is generally a dun brown to grey, with a black or grizzled throat patch. Very similar to Rüppell's Korhaan but is darker on the back and lacks the black line from the centre of the throat to the breast. Easily overlooked in stony veld because it blends so perfectly with the terrain. Usually found in pairs or small groups. *Habitat.* Desert and semi-desert, stony fields and gravel plains, and wheatfields and grassy areas in the southern Cape. Common edemic to the central and south-western regions. *Call.* A rasping, frog-like 'crrok-rrok-rrek-rrek' is given at dawn and dusk. *Afrikaans.* Vaalkorhaan.

236 **Rüppell's Korhaan** *Eupodotis rueppellii* L 58 cm
Confusable only with Karoo Korhaan from which it differs by being generally much paler, almost pinkish grey, and by having a conspicuous black chin which extends in a black line down the front of the neck on to the breast. It blends so perfectly with its background that it is very difficult to detect as it strides across gravel plains. *Habitat.* The gravel plains and semi-scrub desert of Namibia. Common but thinly distributed endemic to the north-west. Usually encountered in small groups of two to three birds. *Call.* Similar to that of Karoo Korhaan. *Afrikaans.* Woestynkorhaan (Damarakorhaan).

231 ▲

236 ▼   230 ▲

235 ▼   232 ▼

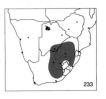

233 Whitebellied Korhaan *Eupodotis cafra* L 52 cm
This pale, sand-coloured korhaan is the only one in the region to have a white belly. Males show a blue neck and breast. Female and imm. similar to female Blackbellied Korhaan but are smaller and lack the black underwing. Usually found in small groups. *Habitat.* Lightly wooded and open grasslands where termite mounds are plentiful. An endemic to the central and eastern regions. *Call.* A crow-like 'krey-krey-krey-krek-krek' is given by the male. *Afrikaans.* Witpenskorhaan.

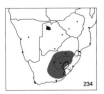

234 Blue Korhaan *Eupodotis caerulescens* L 56 cm
The blue neck and underparts are diagnostic. Female and imm. similar to male but have paler blue underparts. Usually located early in morning when males call. Found normally in small groups of two to three. Very wary, running off at close approach before finally taking flight. *Habitat.* Open grasslands and dry scrub regions. A thinly distributed and uncommon endemic to a small central and south-eastern area. *Call.* A very frog-like 'crrok-crrok-crrok' given only by males and a slightly higher pitched 'crrek-crrek-crrek' given by both sexes. *Afrikaans.* Bloukorhaan.

237 Redcrested Korhaan *Eupodotis ruficrista* L 50 cm
The red crest is rarely seen unless a displaying male is observed, in which case the crest is exposed as an elongation of the nape feathers. Both sexes have black bellies and thin necks. In flight can be distinguished from Blackbellied Korhaan by the lack of white on the upperwings. In its courtship display the male flies straight up, then suddenly tumbles and plummets towards the ground, before gliding off and settling. *Habitat.* Dry thornveld, sometimes in thick bush and grassy areas adjoining thornveld. Common in the northern part of the region. *Call.* The male's call is a 'tic-tic-tic', finishing with a loud, whistling 'chew-chew-chew'. *Afrikaans.* Boskorhaan.

238 Blackbellied Korhaan *Eupodotis melanogaster* L 64 cm
Most likely to be confused with Redcrested Korhaan but is much larger, longer legged, and the male has a black line running from the chin down the front of the neck to the breast. In flight the black underwing contrasts with large white patches on the primaries and distinguishes the female from the female Redcrested and Black korhaans. Male's display flight is spectacular, flying high and then 'parachuting' down with wings held above its head. Usually found singly or in pairs. *Habitat.* Prefers less dense woodland and more open grasslands than Redcrested Korhaan. Common, but usually overlooked, in the north and east. *Call.* A short, sharp 'chikk' followed by a 'pop' is given by the male when sitting on some exposed mound or antheap. *Afrikaans.* Langbeenkorhaan.

239 Black Korhaan *Eupodotis afra* L 52 cm
Of the small korhaans this is the most easily recognized with its black underparts and bright yellow legs. During courtship flight the legs dangle as the bird 'parachutes' slowly to the ground, giving its harsh call. Female distinguished from Blackbellied and Redcrested korhaans by bright yellow legs and conspicuous white forewing patch. *Habitat.* Dry coastal scrub, open grassland and thinly wooded thornveld. Common endemic to the central, western and north-western regions. Absent in the east and north-east. *Call.* Males give a raucous 'kerrrak-kerrrak-kerrrak', both in flight and on the ground. *Afrikaans.* Swartkorhaan.

114

233 ▲   234 ▼

237 ▲   238 ▼

239 ▼

296 Crab Plover *Dromas ardeola* L 38 cm
A large white wader with extensive black patches on its back and wings. The dagger-shaped bill is proportionately very heavy and thick for a bird of this size. Legs are long and greyish. In the imm. the grey in the wings extends up the hind neck to the crown. *Habitat.* Coastal areas and estuaries, and especially muddy mangrove stands, which are rich in crabs. Occurs uncommonly and erratically only on the east coast south to Natal, with one record from the eastern Cape. *Call.* A low, croaking 'kreuk'. *Afrikaans.* Krapvreter.

244 African Black Oystercatcher *Haematopus moquini* L 40 cm
An easily identified large, black wader with a bright orange bill and dull pink legs. Imm. is duller and has a less vivid orange bill. Some ads. show small white patches on the underparts. In flight the feathering is black, showing no wing bars. *Habitat.* This endemic species may be encountered along any shoreline, estuary or lagoon from Port Elizabeth to Namibia, and is a vagrant to coastal Natal. It is common on rocky islands and the adjacent mainland. *Call.* A 'klee-kleeep' call similar to that of European Oystercatcher, and a fast 'peeka, peeka, peeka' alarm call. *Afrikaans.* Swarttobie.

243 European Oystercatcher *Haematopus ostralegus* L 43 cm
Although the same shape as the Black Oystercatcher, this large pied wader has white underparts and a bold white wing bar. The bill is orange, the legs pink. Imm. and many non-breeding ads. show a white throat patch. The broad white wing bar, white rump and base of the tail feathers are visible in flight. The distal half of the tail is black. *Habitat.* Liable to be found on any coast or estuary. This European migrant is a rare vagrant to all coasts of the region, with most records from the Cape. *Call.* A sharp, high-pitched 'klee-kleeep'. *Afrikaans.* Bonttobie.

295 Blackwinged Stilt *Himantopus himantopus* L 38 cm
A large wading bird with very long, red legs and a thin, pointed bill. The wings and back are black; browner in the female and imm. which, in addition, has brown head and neck markings. In breeding plumage the male's crown and nape also become black. In flight the black underwing contrasts with the white underparts, and the long legs trail conspicuously. *Habitat.* Marshes, vleis and flooded ground. When feeding, does not probe, but instead picks from the surface. Common and liable to local movements. *Call.* A harsh, short 'kik-kik', especially when alarmed. *Afrikaans.* Rooipootelsie.

294 Avocet *Recurvirostra avosetta* L 42 cm
An unmistakable white and black wading bird with a long, very thin, upturned bill. In flight the pied pattern is striking, with three black patches on each wing. In the imm. the black is replaced by dusky brown. *Habitat.* Lakes, estuaries, vleis and temporary pools of water. Shallow-water feeder, using a side-to-side bill movement or by up-ending, duck-style. *Call.* A clear 'kooit', also a 'kik-kik' alarm call. *Afrikaans.* Bontelsie.

258 Blacksmith Plover *Vanellus armatus* L 30 cm
This large black, white and grey bird is the easiest plover to identify in our region, and its bold pattern also makes it easily distinguishable in flight. The imm. is a duller version of the ad. with brown feathering replacing the black. *Habitat.* Damp areas at the edges of wetlands, and adjoining grasslands and fields where it feeds. Common in all but the dry west of the region. *Call.* A very vocal species with its loud, ringing 'tink, tink, tink' alarm call. *Afrikaans.* Bontkiewiet.

296 ▲   243 ▼ 295 ▼   244 ▲

258 ▼ 294 ▼

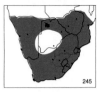

245 Ringed Plover *Charadrius hiaticula* L 16 cm
A small dark plover with a white collar above a blackish brown breast band which is often incomplete. Distinguished from Sand and Mongolian plovers by collar, smaller size, slighter bill and orange-yellow legs. In flight displays an obvious white wing bar. *Habitat.* Coastal and inland wetlands. Common summer migrant. *Call.* A fluty 'tooi'. *Afrikaans.* Ringnekstrandkiewiet.

249 Threebanded Plover *Charadrius tricollaris* L 18 cm
The black double breast band, and conspicuous red eye-ring and base of bill are distinctive. In flight the tail shows a white terminal bar. *Habitat.* Most wetlands but prefers small water bodies. Widespread and common. *Call.* A penetrating, clear 'weet-weet' whistle. *Afrikaans.* Driebandstrandkiewiet.

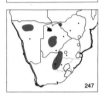

247 Chestnutbanded Plover *Charadrius pallidus* L 15 cm
The smallest, palest plover of the region. Male shows neat black markings on forehead and lores. Ad. has a thin chestnut breast band: in the imm. this is duller and often incomplete. Differentiated from larger Whitefronted Plover by the breast band if present, and lack of white collar. *Habitat.* Salt pans and coastal wetlands, from the eastern Cape to Namibia. *Call.* A single 'tooit'. *Afrikaans.* Rooibandstrandkiewiet.

248 Kittlitz's Plover *Charadrius pecuarius* L 16 cm
The black forehead line that extends behind the eye on to the nape is distinctive. The breast is creamy buff and there is a dark shoulder patch. Imm. distinguished from Whitefronted Plover by buffy nape and dark shoulder. *Habitat.* Common throughout in dried muddy or short grassy areas near water. *Call.* A short, clipped trill 'kittip'. *Afrikaans.* Geelborsstrandkiewiet.

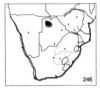

246 Whitefronted Plover *Charadrius marginatus* L 18 cm
A very pale plover which resembles the Chestnutbanded Plover but is slightly larger, has a white collar and lacks a complete breast band. Paler than Kittlitz's Plover and lacks the dark head markings. It is also lighter coloured, clearer breasted and smaller than Lesser, Sand and Ringed plovers. *Habitat.* Sandy beaches, muddy coastal areas and inland rivers. *Call.* A clear 'wiiit', but has a 'tukut' alarm call. *Afrikaans.* Vaalstrandkiewiet.

252 Caspian Plover *Charadrius asiaticus* L 22 cm
In breeding plumage differs from Mongolian and Sand plovers by the black lower border to its chestnut breast band. In all other plumages it has a complete (or virtually complete) grey-brown wash across the breast. Bill is small and thin. In flight lacks distinct white areas on the upperparts. Distinguished from winter plumage Lesser Golden Plover by plain wings, and smaller size. *Habitat.* Summer migrant to sparse, grassy areas and wetlands. Erratically distributed. *Call.* A clear whistle, 'tooeet'. *Afrikaans.* Asiatiese Strandkiewiet.

250 Mongolian Plover *Charadrius mongolus* L 19 cm
Resembles the Sand Plover but is shorter legged, smaller bodied and has a shorter, less robust bill. Both ad. Whitefronted and Kittlitz's plovers are smaller and have a pale collar. When breeding, the rufous breast band becomes more extensive. *Habitat.* Coastal wetlands. Summer visitor to coastal Natal, vagrant to the Cape and Namibia. *Call.* 'Chittick'. *Afrikaans.* Mongoolse Strandkiewiet.

251 Sand Plover *Charadrius leschenaultii* L 22 cm
Very similar to Mongolian Plover but stands taller, has a bigger body and a longer, more robust bill. Differs from Whitefronted Plover by its larger size and by lacking the white collar. Caspian Plover is smaller billed and lacks extensive white on the outer tail feathers. When breeding, the brown shoulder patches become rufous and extend across the breast. *Habitat.* Coastal wetlands, in the east. *Call.* A short trill. *Afrikaans.* Grootstrandkiewiet.

118

245 ▲   247 ▼   248 ▼   249 ▲

246 ▲   250 ▼   251 ▼   252 ▲

272 Curlew Sandpiper *Calidris ferruginea* L 19 cm
The only small, plain wader with an obviously decurved bill. Dunlin is shorter
legged, smaller, has a very slightly decurved bill, and a dark line down the
rump. In flight, shows a squarish white rump. In breeding plumage, the rump
may become finely barred and the underparts rust-coloured. *Habitat.* Very
common, gregarious summer visitor to virtually any water body. Regularly
overwinters. *Call.* A short trill, 'chirrup'. *Afrikaans.* Krombekstrandloper.

273 Dunlin *Calidris alpina* L 18 cm
The Dunlin's bill is less decurved than that of the Curlew Sandpiper, from
which it also differs by its smaller size and by its dark, not white rump. The
Broadbilled Sandpiper is smaller, has shorter, paler legs (Dunlin's are black),
striped head markings and a bill with a slightly flattened tip. In breeding and
transitional plumage, shows a black patch on the belly. *Habitat.* The edges of
wetlands. Rare vagrant, recorded only from the Cape and Transvaal.
Call. A weak 'treep'. *Afrikaans.* Bontstrandloper.

283 Broadbilled Sandpiper *Limicola falcinellus* L 17 cm
A small, short-legged wader distinguished from Little Stint by its bill which
droops slightly at its flattened tip. The legs are grey, sometimes greenish, but
never black. There is a dark shoulder patch which is less pronounced than in
the Sanderling. Its head pattern is diagnostic: a whitish, double eyebrow stripe
which gives the head a striped appearance. In breeding plumage the
upperparts become blackish with narrow, pale buffy feather margins.
Habitat. Vleis, lakes and estuaries. Uncommon but regular summer visitor.
Call. A fairly low-pitched, short trill, 'drrrt'. *Afrikaans.* Breëbekstrandloper.

278 Baird's Sandpiper *Calidris bairdii* L 17 cm
Larger than the stints. Distinguished by its markedly scaled, essentially brown,
not grey upperparts, and by the light streaking encompassing the breast. The
legs are black. Pectoral Sandpiper is larger and has greenish legs. Sanderling
attaining breeding plumage is similar but is bigger and has a pronounced,
paler wing bar. Ruff also has scaled wing coverts but is much bigger, has
white sides to the rump and does not have black legs. *Habitat.* Edges of
wetlands. Very rare vagrant, with one specimen recorded from Namibia.
Call. A short trill, 'kreep'. *Afrikaans.* Bairdse Strandloper.

277 Whiterumped Sandpiper *Calidris fuscicollis* L 17 cm
This small, rather plain sandpiper has a white rump like Curlew Sandpiper, but
is smaller and has a shorter, straighter bill. Knot has a pale but not white rump
and is much larger. *Habitat.* Wetlands. Rare vagrant: only three records, from
the Cape and Namibia. *Call.* A thin 'jeep'. *Afrikaans.* Witrugstrandloper.

282 Buffbreasted Sandpiper *Tryngites subruficollis* L 18 cm
The only sandpiper with completely buff-coloured underparts. It has a rather
short bill, yellow legs, and a white eye-ring visible at close quarters. In flight it
shows neither a wing bar nor a distinct rump or tail pattern, and has a white
underwing. The imm. has a white belly. *Habitat.* Frequents short grassy areas
and wetlands. Rare vagrant with one sighting from Natal. *Call.* Not recorded in
our region. *Afrikaans.* Taanborsstrandloper.

279 Pectoral Sandpiper *Calidris melanotos* L 19 cm
The abrupt definition between the streaked breast and the white underparts is
distinctive. Larger than the stints, it is longer necked with a slightly dark-
capped appearance, a reddish base to the bill and darker upperparts. The
legs are yellowish. Baird's Sandpiper has a black bill and legs, and the much
larger Ruff lacks the breast marking. *Habitat.* Margins of smaller wetlands
which have some short vegetation. A rare but widely recorded summer visitor.
Call. A low trill, 'prrrt'. *Afrikaans.* Geelpootstrandloper.

120

283 ▲   273 ▼   278 ▼

277 ▲   272 ▼   279 ▼   282 ▲

253 **Lesser Golden Plover** *Pluvialis dominica* L 24 cm
Distinguished from Grey Plover by being smaller, having warm buff speckling on the wing coverts and back and, in flight, by the lack of black axillaries and whitish rump. Imm. Caspian Plover resembles this species but is smaller, paler and has larger buff speckling on the back. May assume summer plumage before migrating, in which case it resembles the breeding Grey Plover, except that the wing speckling is golden, not white. *Habitat.* Wetland habitats and their surrounding exposed areas. A rare vagrant recorded throughout the region but mostly on the coast. *Call.* A single- or double-note, high-pitched whistle 'oodle-oo'. *Afrikaans.* Goue Strandkiewiet.

254 **Grey Plover** *Pluvialis squatarola* L 30 cm
In flight its black axillaries and pale rump distinguish it from Lesser Golden Plover. At rest shows grey, not buff speckling on the back and wing coverts. The Knot is smaller, has shorter legs, a longer bill and plainer grey upperparts. In summer plumage shows whitish speckling on the back and wings, not golden as in Lesser Golden Plover. *Habitat.* Bays, lagoons and estuaries where there are expanses of mud. Occasionally seen inland. Common summer migrant, with some birds overwintering. *Call.* A clear 'tluui', lower in pitch in the middle. *Afrikaans.* Grysstrandkiewiet.

262 **Turnstone** *Arenaria interpres* L 23 cm
This distinctive wader has a short, very slightly upturned bill, and orange legs. In winter plumage, the upperparts are blackish with irregular dark markings on the front and sides of the breast. Breeding plumage, in which the head and neck achieve a crisp black and white pattern and the wings take on a warm brown colouring, is often attained before departure in March/April. In flight in all plumages, the upperparts show a distinctive dark and light pattern. *Habitat.* Stony shorelines, where it uses its bill to overturn stones and debris in its search for food. Common in wetlands and along coasts throughout. *Call.* A hard 'kttuck', especially in flight. *Afrikaans.* Steenloper.

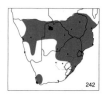

242 **Painted Snipe** *Rostratula benghalensis* L 24-28 cm
The female is more striking than the male, with a chestnut neck and breast. Distinguished from Ethiopian and Great snipes by breast pattern, longer legs and shorter, decurved bill. Has a laboured flight action. The imm. is like the male. *Habitat.* Skulks amongst reeds in marshes and on the edges of lakes, vleis and dams. Occurs very erratically throughout, being nowhere common. *Call.* The male utters a trill, the female makes a hissing sound and has a croaking 'kook' call. *Afrikaans.* Goudsnip.

285 **Great Snipe** *Gallinago media* L 35 cm
Likely to be confused with the Ethiopian Snipe which has white on the belly, not the buffy wash overlaid with dark barring of this species. It differs further by having a shorter bill and, at close quarters, the white tail tip and white outer tail feathers can be distinguished; those of the Ethiopian Snipe are barred. Flight is heavy and direct on rounded wings. *Habitat.* This summer migrant occurs in vegetation of marshes and on the edges of water bodies. Previously more abundant, it is now rare in the north and east. *Call.* Generally silent but may croak when put to flight. *Afrikaans.* Dubbelsnip.

286 **Ethiopian Snipe** *Gallinago nigripennis* L 32 cm
The long straight bill and striped head markings distinguish it from all but the Great Snipe, from which it is separated by its longer bill, barred, not white outer tail feathers and a white, not buff belly with dark barring restricted to the flanks. Differs too by having faster, jerkier flight. *Habitat.* A common resident of the vegetation surrounding wetlands. *Call.* A grunting sound when put to flight. In the breeding season, its fanned tail feathers produce a drumming sound during aerial display. *Afrikaans.* Afrikaanse Snip.

122

253 ▲

262 ▼ 242 ▼

254 ▲

285 ▼ 286 ▼

255 **Crowned Plover** *Vanellus coronatus* L 30 cm
Unmistakable with its black cap interrupted by a white 'halo'. Legs and basal part of bill bright reddish. Imm. much like ad. but less vividly marked. Even for a plover it has a very upright stance. *Habitat.* Has no particular affinity for water, preferring drier fields, parks, and grazed and agricultural land. Congregates in small flocks, especially when not breeding. A common resident throughout the region. *Call.* An extremely noisy species, uttering a loud, grating 'kreeep' call, day or night. *Afrikaans.* Kroonkiewiet.

257 **Blackwinged Plover** *Vanellus melanopterus* L 27 cm
Similar to Lesser Blackwinged Plover but is larger, shows more white on the forehead, has a broader black border separating the breast from the belly and, at close quarters, its red, not brown eyelids can be seen. In flight the secondaries are white, tipped with black, and the underwing is pure white. The legs are dark red. *Habitat.* Exposed grassy upland areas in the eastern sector of the region as far south as Mossel Bay. An uncommon summer migrant and local resident, with movements from mountains to the coast. *Call.* A high-pitched, ringing call consisting of repeated 'kiya-kiya' phrases. *Afrikaans.* Grootswartvlerkkiewiet.

256 **Lesser Blackwinged Plover** *Vanellus lugubris* L 22 cm
Closely resembles Blackwinged Plover but is smaller, has a narrow black border to the breast, less white on the forehead and has brown, not red eyelids. In flight the upperwing shows completely white secondaries. The underwing is greyish, not pure white. *Habitat.* Dry, open grassy areas. Occurs very erratically in the eastern lowveld and northern Natal, and is a vagrant to Zimbabwe. Uncommon and thinly distributed. *Call.* A clear, double-note 'tee-yoo, tee-yoo'. *Afrikaans.* Kleinswartvlerkkiewiet.

260 **Wattled Plover** *Vanellus senegallus* L 35 cm
The largest plover of the region. Has smaller yellow wattles than Whitecrowned Plover, from which it also differs by having a dark breast bordered by a thin black line on the belly. The white on the head is restricted to the forecrown and forehead, unlike the Whitecrowned Plover in which the white stripe extends beyond the crown. In flight this species shows more black on the trailing edge of the wing than does the Whitecrowned Plover. *Habitat.* This fairly common resident plover frequents the damper borders of wetlands. Found in the northern and north-eastern sectors of the region with individuals wandering further south. *Call.* A high-pitched, ringing 'keep-keep'. *Afrikaans.* Lelkiewiet.

259 **Whitecrowned Plover** *Vanellus albiceps* L 30 cm
This large plover with its distinctive pendulous, yellow wattles might be confused with the Wattled Plover which has a dark, not white breast. The Whitecrowned Plover is also distinguished by a white stripe running from the forehead to the nape. In flight the black wing tip with the white trailing edge is visible. The imm. is like the ad. *Habitat.* In the northern part of the region, a fairly common resident of the major river systems, where it frequents sand bars and exposed banks. *Call.* Typical of large plovers, its call is a repeated, ringing 'peek-peek'. *Afrikaans.* Witkopkiewiet.

261 **Longtoed Plover (Whitewinged Plover)** *Vanellus crassirostris* L 30 cm
The only plover in the region to have a white face and throat. The black nape extends down the sides of the neck to form a broad breast band. In flight, it looks very striking with its wings being all white except for the black outer primaries. *Habitat.* It has jacana-like habits of foraging on floating vegetation, utilizing its long toes to distribute its weight. Apart from southern vagrants, it is restricted to the most northern and north-eastern sectors of the region where it occurs regularly in suitable habitat. *Call.* A repeated high-pitched 'pink-pink'. *Afrikaans.* Witvlerkkiewiet.

255 ▲

256 ▼ 260 ▼

257 ▲

259 ▼ 261 ▼

274 **Little Stint** *Calidris minuta* L 14 cm
Very similar to Rednecked Stint but the latter has a shorter, stubbier bill. Legs are black. In breeding plumage, lacks the rufous throat and neck of Rednecked Stint. Temminck's Stint has olive, not black legs, plainer upperparts and white, not grey outer tail feathers. Longtoed Stint has darker upperparts, a thinner bill and greenish yellow legs. Broadbilled Sandpiper and Dunlin have bills which droop very slightly at the tip, and the former shows a dark shoulder when at rest. *Habitat.* Common migrant throughout, preferring larger lakes, bays and estuaries with extensive muddy areas. *Call.* A repeated 'peep'. *Afrikaans.* Kleinstrandloper.

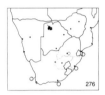

276 **Rednecked Stint** *Calidris ruficollis* L 14 cm
Extremely difficult to distinguish from Little Stint except in breeding plumage when the throat, neck and head become rufous. In all other plumages, Rednecked Stint's bill is shorter and stubbier, and its crown and nape are more clearly streaked. In non-breeding plumage it is very pale, with the wing coverts and back feathers having only a thin black line down their centres. Separated from Temminck's and Longtoed stints by its black, not greenish or yellow legs. *Habitat.* Usually occurs with Little Stint or Curlew Sandpiper at vleis, dams, estuaries or bays. A very uncommon summer visitor, it has been recorded from the Cape, Natal and Transvaal. However, may be more widespread than realized. *Call.* A double-syllable 'tirriw', similar to Little Stint's call. *Afrikaans.* Rooinekstrandloper.

280 **Temminck's Stint** *Calidris temminckii* L 14 cm
Distinguished by greenish legs, pure white outer tail feathers and uniform grey-brown upperparts. Longtoed Stint generally has yellow legs and is much darker and browner above. *Habitat.* Water margins with short grass or reeds. Could occur on any vlei, dam or estuary. Rare vagrant: one record from the Transvaal. *Call.* A shrill 'prrrrtt'. *Afrikaans.* Temminckse Strandloper.

284 **Ruff** *Philomachus pugnax* L ♂ 30 cm; ♀ 24 cm
The male is bigger than the female. The obvious scaled pattern of the upperparts is distinctive. The black bill may show an orange base, and the colour of the legs is highly variable. Similar to Redshank which lacks the scaled upperparts and has striking white secondaries. The white underparts distinguish it from the Buffbreasted Sandpiper. In flight, the white oval patch on either side of the rump is diagnostic. The male may be seen in partial breeding plumage: a white head and neck ruffs of various colours. *Habitat.* Vleis, lakes, estuaries and adjacent grassy areas. A summer migrant, common throughout. *Call.* Silent in our region. *Afrikaans.* Kemphaan.

271 **Knot** *Calidris canutus* L 25 cm
This short-legged, dumpy, rather plain wader differs from the smaller Curlew Sandpiper by its straight bill. It is much smaller than the Grey Plover which has speckled, not uniformly coloured back, and a small bill. In flight a pale wing bar is apparent and the rump is pale grey, not white. *Habitat.* Gregarious wader of estuaries and bays, most common on the west coast. Occasionally occurs inland. *Call.* A nondescript 'knut'. *Afrikaans.* Knoet.

275 **Longtoed Stint** *Calidris subminuta* L 14 cm
Separated from Little and Rednecked stints by its darker brown upperparts, yellow, not black legs and more slender bill. Some Temminck's Stints have yellow legs but they have plainer, paler upperparts and white, not greyish edges to the tail. Only when very close can the long toes be seen. When alarmed stands very erect and, when flushed, may 'tower' as opposed to flying horizontally away. *Habitat.* Liable to occur on any water body. Recorded from coastal Moçambique and the Cape. Very rare. *Call.* Not recorded in our region. *Afrikaans.* Langtoonstrandloper.

274▲ 280▼ 284▼ 276▲

271▼ 275▼

263 Terek Sandpiper *Xenus cinereus* L 23 cm
The only small wader with a long, upturned, dark brown bill with an orange base. The rump is pale and the short, bandy legs are orange-yellow. Similar to Curlew Sandpiper in size but has shorter legs. At a distance it appears fairly pale with a dark shoulder, and the white trailing edge to the secondaries is clearly visible in flight. *Habitat.* Muddy estuaries and bays, especially near mangroves. A summer visitor to all coasts but more common in the north and east. *Call.* A series of fluty, uniformly low-pitched 'du-du-du' notes. *Afrikaans.* Terekruiter.

264 Common Sandpiper *Tringa hypoleucos* L 19 cm
Normally shows an obvious white shoulder in front of the closed wing. Wood Sandpiper is larger, longer legged and has pale spotting on the upperparts. Green Sandpiper is larger and has a white rump and black underwing. Flight consists of bursts on slightly bowed wings, interspersed with short glides. Shows a prominent pale wing bar and barred sides to the dark tail. At rest it bobs its tail. *Habitat.* This common summer visitor may be found in a wide range of wetland habitats throughout the region. *Call.* A very shrill 'ti-ti-ti', higher and thinner than that of Wood Sandpiper. *Afrikaans.* Gewone Ruiter.

265 Green Sandpiper *Tringa ochropus* L 23 cm
Resembles Wood and Common sandpipers but is larger, has darker upperparts which contrast with its square white rump, a black underwing and a more prominent eyebrow stripe. Greenshank and Marsh Sandpiper are very much paler and lack the black underwing. *Habitat.* Wetlands, where it tends to use vegetation for cover. A rare summer visitor to the north and east. *Call.* A three-note whistle 'tew-a-tew'. *Afrikaans.* Witgatruiter.

267 Spotted Redshank *Tringa erythropus* L 32 cm
Similar to the smaller Redshank but is paler, longer billed and lacks the white trailing edge to the secondaries. Legs and base of bill are dull reddish, which distinguishes it from other sandpipers except the Ruff. Ruff lacks the triangular white rump. In breeding plumage this species becomes black, finely spotted with white. *Habitat.* The edges of vleis, lakes, estuaries, bays and lagoons. Rare vagrant with records from Zimbabwe, Moçambique and Natal. *Call.* A clear, double-note 'tu wik'. *Afrikaans.* Gevlekte Rooipootruiter.

268 Redshank *Tringa totanus* L 25 cm
The red base to the bill and the orange-red legs differentiate this from all other waders except the Ruff and Spotted Redshank. The Ruff (which sometimes has orange legs and base to the bill) lacks the triangular white rump; the Spotted Redshank is paler, longer billed, and both those species lack the striking white trailing edge to the secondaries. At rest the Redshank's wings and back appear plain brown, unlike the Ruff's scaled and mottled upperparts. *Habitat.* The margins of vleis, dams, estuaries and lagoons. A rare vagrant from Europe to the west coast, with scattered records from the eastern sector. *Call.* 'Tiw-hu-hu', the first syllable being musically a tone above the others. *Afrikaans.* Rooipootruiter.

266 Wood Sandpiper *Tringa glareola* L 20 cm
Closely resembles and, in size, is intermediate between Common and Green sandpipers. Distinguished from Green Sandpiper by paler upperparts and underwing, less streaking on the breast and less prominent eye-stripe. Lacks the white shoulder and wing bar of Common Sandpiper and has a different flight action. Much darker and browner than Greenshank or Marsh Sandpiper. *Habitat.* Dams, vleis, bays, estuaries and even puddles. This summer visitor is common throughout the region, occurring singly or in small flocks. *Call.* A very vocal species with a high-pitched, slightly descending 'chiff-iff-iff'. *Afrikaans.* Bosruiter.

263 ▲   265 ▼ 267 ▼   264 ▲

268 ▼ 266 ▼

269 Marsh Sandpiper *Tringa stagnatilis* L 23 cm
This pale grey sandpiper resembles the Greenshank but has a thinner, straight black bill, slighter body build and proportionately longer legs. Winter plumage Wilson's Phalarope is smaller, has yellow, not grey-green legs and has a dark line through the eye. *Habitat.* Estuaries and larger inland water bodies. Common throughout. *Call.* Usually just a single 'tchuck', which is not high-pitched or shrill. *Afrikaans.* Moerasruiter.

270 Greenshank *Tringa nebularia* L 32 cm
Pale grey like the Marsh Sandpiper but is larger and has an upturned black bill with a grey base. Bartailed Godwit is bigger, browner and has an upturned bill with a pinkish base. Greenshank's upturned bill differentiates it from all other waders except the smaller, orange-legged Terek Sandpiper. *Habitat.* Common summer migrant on wetlands throughout the region. *Call.* A loud, rasping 'chew-chew-chew'. *Afrikaans.* Groenpootruiter.

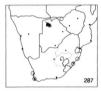

287 Blacktailed Godwit *Limosa limosa* L 40 cm
This large wader resembles the Bartailed Godwit but at rest differs by having a greyer, unmottled back and an almost straight, pink-based bill. In flight it is distinguished by its broad white wing bars, and white tail with a black tip. *Habitat.* Larger lakes, estuaries and bays. Rare migrant which occurs throughout except in Moçambique. *Call.* 'Weeka-weeka', especially in flight. *Afrikaans.* Swartstertgriet.

288 Bartailed Godwit *Limosa lapponica* L 38 cm
At rest resembles Blacktailed Godwit except that the bill is shorter and slightly upturned, and the upperparts are browner and more distinctly marked. In flight this species lacks the Blacktailed's white wing bars and shows light bars, not a black band on the tail. *Habitat.* Estuaries, bays and muddy-edged lakes. Summer migrant common on all coasts. *Call.* Generally silent but does utter a 'wik-wik' call. *Afrikaans.* Bandstertgriet.

289 Curlew *Numenius arquata* L 60 cm
A very large wader with an extremely long decurved bill which is proportionately longer than the Whimbrel's. It is overall paler and lacks the head stripes of that species. In flight it shows a conspicuous white rump that extends up the back as a white triangle. *Habitat.* Prefers large estuaries and bays although some birds do venture inland. Occurs on all coasts, but is now rare in the east. *Call.* A loud 'cur-lew'. *Afrikaans.* Grootwulp.

290 Whimbrel *Numenius phaeopus* L 43 cm
Apart from the Curlew, this is the only large wader with a decurved bill. Whimbrel is smaller, shorter-billed, and darker than Curlew, and has dark stripes on the head and through the eye. In flight the tail is barred dusky brown. *Habitat.* Rarely seen outside its habitat of estuaries and bays. Small numbers of this common summer migrant regularly overwinter. *Call.* An evenly pitched, bubbling call of about seven syllables. *Afrikaans.* Kleinwulp.

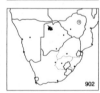

902 Lesser Yellowlegs *Tringa flavipes* L 28 cm
Closely resembles Wood Sandpiper, but is larger and has a more streaked breast. The legs are yellower than any other similar-sized wader of the region. *Habitat.* Small vleis to large bays. Forages in mud and waterside vegetation. Extremely rare vagrant: one record from Zimbabwe and one from the Cape. *Call.* A repeated ringing 'chew', similar to that of Wood Sandpiper.

906 Greater Yellowlegs *Tringa melanoleuca* L 35 cm
Smaller than the Greenshank, it shows yellow, not green legs, and has a square white rump which does not extend up the back. *Habitat.* Freshwater dams and estuaries. Rare vagrant, with one record from the Cape. *Call.* Not recorded in Africa.

130

◀269

288▼ 287▼ 902▼

290▼ 906▼

270▼ 289▼

293 **Wilson's Phalarope** *Phalaropus tricolor* L 22 cm
Distinguished from the other, smaller phalaropes by its longer, thinner bill, paler upperparts and yellow, not dark legs. In flight it differs by having a white rump, grey tail and no wing bar. In breeding plumage a distinct dark stripe runs through the eye and down the side of the neck, and the sides of the breast and mantle attain a rufous wash. *Habitat.* Very rare vagrant to Cape and Natal estuaries. *Call.* Not recorded in our region. *Afrikaans.* Bontfraiingpoot.

292 **Rednecked Phalarope** *Phalaropus lobatus* L 16 cm
Resembles Grey Phalarope but has a darker grey back streaked with white, and a longer, thinner, all-black bill. Wilson's Phalarope is much paler with a longer, thinner bill and yellow, not black legs. In flight the upperparts are darker than Grey Phalarope's and the white rump of Wilson's Phalarope is lacking. In breeding plumage it acquires a small chestnut gorget on the upper neck. *Habitat.* Any quiet body of water. A vagrant to the Cape and Natal. *Call.* A low 'tchuck' when put to flight. *Afrikaans.* Rooihalsfraiingpoot.

291 **Grey Phalarope** *Phalaropus fulicarius* L 18 cm
Similar to Rednecked Phalarope but is more uniformly pale grey above, and has a shorter, thicker, sometimes yellow-based bill. Wilson's Phalarope is larger, has a longer, thinner bill, a less obvious dark line through the eye and, in flight, has a white, not dark rump and shows no wing bar. In breeding plumage, the chestnut underparts are diagnostic. *Habitat.* A summer visitor to the open seas of the western Cape. A vagrant to the coast and mainland, it also occurs inland. *Call.* A soft, low 'wiit'. *Afrikaans.* Grysfraiingpoot.

281 **Sanderling** *Calidris alba* L 19 cm
In non-breeding plumage the palest sandpiper of the region. It has a rather short, stubby bill and a dark shoulder. Similar to the dark-shouldered Broadbilled Sandpiper but is larger, heavier billed and lacks head stripes. Individuals in breeding plumage have wing and back feathers black with rufous centres. In flight shows a distinct white wing bar. *Habitat.* Sandy beaches, estuaries and bays. Common on the coast with individuals occurring inland. *Call.* A single, decisive 'wick'. *Afrikaans.* Drietoonstrandloper.

305 **Blackwinged Pratincole** *Glareola nordmanni* L 27 cm
At rest appears darker than Redwinged Pratincole. In flight the white rump is in sharp contrast with the rest of the upperparts, a black, not rufous underwing can be seen, and it lacks the white trailing edge to its secondaries. Imm. is drabber, without the black throat lines. *Habitat.* A summer migrant which prefers dry habitats. Partly nomadic, being least common to the west and east. *Call.* A single, often-repeated 'pik'. *Afrikaans.* Swartvlerksprinkaanvoël.

304 **Redwinged Pratincole** *Glareola pratincola* L 27 cm
The ad. has a buff throat edged with a thin black line. Similar to Blackwinged Pratincole but has dark rufous, not black axillaries and appears generally paler brown. The flight is light and graceful, and a white rump and forked tail are displayed. Imm. lacks the clearly defined throat markings of the ad. *Habitat.* The edges of wetlands, especially where there are sandbanks. Localized areas in the extreme north and east, south to Natal. *Call.* 'Kik-kik', especially in flight. *Afrikaans.* Rooivlerksprinkaanvoël.

306 **Rock Pratincole (Whitecollared Pratincole)** *Glareola nuchalis* L 18 cm
Much smaller than Redwinged Pratincole. A diagnostic white line extends down from the eye across the lower nape to form a collar. The legs and the base of the bill are red. The rump shows up starkly white in flight. Imm. is a dull version of the ad. and has darker legs. *Habitat.* A resident of the Zambezi River where it frequents stretches of water with exposed flat rocks and sandbars. *Call.* A loud, plover-like 'kik-kik'. *Afrikaans.* Withalssprinkaanvoël.

132

281 ▲   292 ▼ 305 ▼   304 ▼   293 ▲

291 ▼   306 ▼

299 Burchell's Courser *Cursorius rufus* L 23 cm
A plain, buff-grey courser with a black and white line extending back from the eye. Separated from Temminck's Courser by its blue-grey, not rufous crown and nape, and the black bar, as opposed to patch, on the belly. In flight there is a substantial white bar on the secondaries and a white tip to the outer tail; Temminck's Courser has a thin bar on the secondaries and extensive white on the outer tail feathers. The imm. is mottled above. *Habitat.* Dry, sparsely grassed areas in the west, south, and centre. Erratic and unpredictable distribution. *Call.* A harsh, repeated 'wark'. *Afrikaans.* Bloukopdrawwertjie.

300 Temminck's Courser *Cursorius temminckii* L 20 cm
Most resembles Burchell's Courser from which it is distinguished by being generally more grey-brown above, having a rufous, not grey hind crown and a black patch, not line, on its lower belly. Above this black patch is an irregular rufous area. In flight, the outer tail shows up white, and the underwing is black with a contrasting white trailing edge. The imm. is duller and has lightly speckled upperparts. *Habitat.* A nomadic species which occurs erratically, especially in the north and east of the region, mostly absent from the extreme south and west. Prefers dry, sparsely grassed areas. *Call.* A piercing 'keer-keer'. *Afrikaans.* Trekdrawwertjie.

301 Doublebanded Courser *Rhinoptilus africanus* L 22 cm
The two narrow black bands ringing the upper breast are diagnostic. This species has the plainest head of the coursers. Wing and back feathers are marked with dark centres and contrasting pale fringes. In flight the uppertail coverts show up white. *Habitat.* A resident of dry open areas, including desert, but absent from the east and north-east. *Call.* A 'weqk' whistle and repeated 'kee-kee' notes. *Afrikaans.* Dubbelbanddrawwertjie.

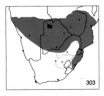

303 Bronzewinged Courser *Rhinoptilus chalcopterus* L 25 cm
A fairly dark brown and white courser with a broad dusky band across the breast and lower neck. The belly is white and there is an irregular white area on the throat and upper neck. In flight, the white uppertail coverts and wing bars contrast with the dark upperparts. *Habitat.* Dry, grassy, lightly wooded areas in all but the dry southern and western regions. Uncommon, subject to local irruptions. *Call.* A ringing 'ki-kooi'. *Afrikaans.* Bronsvlerkdrawwertjie.

302 Threebanded Courser *Rhinoptilus cinctus* L 28 cm
The largest of our coursers, this species has three bands (the lowest of which is rufous) across the neck and breast. The white forehead extends back in a stripe over the eye towards the nape. In flight, no white shows on the upperparts. The imm. resembles the ad. but the breast bands are less distinct. *Habitat.* Dry woodland. Occurs only in the extreme north, especially Zimbabwe. *Call.* A repeated 'kika-kika-kika'. *Afrikaans.* Driebanddrawwertjie.

298 Water Dikkop *Burhinus vermiculatus* L 40 cm
At rest likely to be confused only with the Spotted Dikkop but is smaller, has a white wing bar, unspotted upperparts and a white line above and below the eye. The white wing bar distinguishes it in flight. *Habitat.* Wetlands. Common in suitable habitat. *Call.* 'Ti-ti-tee-teee-tooo', retarding and dropping in pitch at the end. *Afrikaans.* Waterdikkop.

297 Spotted Dikkop (Cape Dikkop) *Burhinus capensis* L 44 cm
Larger and darker than the Water Dikkop and with obviously spotted upperparts. Unlike the Water Dikkop, it shows no wing bar, displaying instead two small white patches on each upperwing when in flight. Lacks the white line above and below the eye seen in the Water Dikkop. *Habitat.* Dry sparse bush and grazed areas. Common throughout the drier regions. *Call.* 'Whiw-whiw-whiw', especially at night. *Afrikaans.* Dikkop.

134

300▲ 299▼ 303▼ 301▼ 302▲

298▼ 297▼

307 **Arctic Skua** *Stercorarius parasiticus* L 46 cm
Ad. plumages are very variable: light, dark and intermediate phases occur. Pomarine Skua is bulkier, broader winged, larger billed and the central tail feathers, if present, are spoon-shaped, not pointed. Longtailed Skua is greyer, slimmer, has narrower wings and much longer, pointed central tail feathers, if present. Imm. unlikely to be distinguished from imm. Longtailed Skua unless seen together, when the larger size and heavier proportions of this species are evident. *Habitat.* Prefers inshore waters, where it parasitizes terns and gulls. Common off all our coasts during summer. *Call.* Normally silent in our region. *Afrikaans.* Arktiese Roofmeeu.

309 **Pomarine Skua** *Stercorarius pomarinus* L 50 cm
Most common in the light phase when the ad. is identified by spoon-shaped, not pointed central tail feathers, although these feathers are often worn or absent. In comparison with the Arctic and Longtailed skuas, this species is more heavily built, longer billed, broader winged, and has more white in the wing. Imm. is more heavily barred on the upper- and undertail coverts than Arctic Skua. *Habitat.* Southern African seas where it harries terns, gulls and other seabirds. Although not as common as the Arctic Skua, especially inshore, it is a regular summer visitor. *Call.* Silent in our region. *Afrikaans.* Knopstertroofmeeu.

308 **Longtailed Skua** *Stercorarius longicaudus* L 50 cm
Ad. in breeding plumage has elongated central tail feathers. The dark phase is extremely rare. Pale phase bird has slimmer body, wings and bill, colder grey-brown plumage and a smaller white patch at the base of the primaries than Arctic or Pomarine skuas. Imm. virtually indistinguishable from imm. Arctic Skua except by its smaller size and more tern-like proportions. *Habitat.* More of a scavenger than other skuas, it attends trawlers in the open sea. A regular summer visitor to Cape and Namibian waters but a vagrant to the east coast. *Call.* Silent in our region. *Afrikaans.* Langstertroofmeeu.

310 **Subantarctic Skua** *Catharacta antarctica* L 60 cm
The largest skua, this species is heavy bodied, broad winged, and dark brown in colour. The distinct white patch at the base of the primaries distinguishes it from the imm. Pomarine Skua which has a less obvious wing patch. Differentiated from South Polar Skua by uniform, not contrasting wings and underparts. Imm. Kelp Gull has pale, not concolorous uppertail coverts. *Habitat.* Southern African seas with greatest numbers occurring during the winter months. An aggressive scavenger, it preys on other seabirds. *Call.* A soft 'wek-wek' and a loud 'yap-yap'. *Afrikaans.* Bruinroofmeeu.

311 **South Polar Skua** *Catharacta maccormicki* L 53 cm
Variable in plumage pattern but dark upperparts and underwing always contrast with paler head, nape and underparts. Both this and the Subantarctic Skua have a white patch near the wing tip, but the latter species lacks the contrasting plumage coloration. Imm. is darker than the ad. but also shows a diagnostic pale collar. Not confusable with imm. Kelp Gull as the latter has a barred rump. *Habitat.* The open sea where it scavenges and harries other seabirds. A rare vagrant to Cape waters. *Call.* Not recorded within our region. *Afrikaans.* Suidpoolroofmeeu.

311▲ 308▼ 307◢

309▼ 310▼

314 Herring Gull *Larus argentatus* L 55 cm ☐
The subspecies, *L.a. hueglini,* which has occurred in Natal, is now considered a race of the Lesser Blackbacked Gull. *Afrikaans.* Haringmeeu.

313 Lesser Blackbacked Gull *Larus fuscus* L 55 cm
At all ages distinguished from Kelp Gull by slightly smaller size, more attenuated appearance with the wings projecting well beyond the tail when at rest, and less robust bill. Compared with the Kelp Gull, the ad. has rich yellow, not olive legs; pale yellow, not dark eyes; and dark grey, not black upperwing and back. The imm. is separated by having flesh-coloured, not brown legs and generally by its slender proportions and shape. Imm. might be confused with the large skuas but it lacks their white wing patches and has a barred, not brown rump. *Habitat.* A scavenger at lakes, bays and along coasts. The only large, dark-backed gull to be found well inland. Regular visitor to the Natal coast. *Call.* A typical large gull 'kow-kow' and shorter 'kop' call. *Afrikaans.* Kleinswartrugmeeu.

312 Kelp Gull (Southern Blackbacked Gull) *Larus dominicanus* L 60 cm
The largest gull of the region. Ad. differs from ad. Lesser Blackbacked Gull by having dark, not pale yellow eyes, and olive, not yellow legs. Sub-ad. birds are distinguished by being bigger and more robust than the Lesser Blackbacked, and by having brown, not pink legs. Imm. resembles Subantarctic and South Polar skuas but lacks the white wing patches and has a barred rump. *Habitat.* Forages along coasts, estuaries, inshore waters and may venture a few kilometres inland. Resident on all our coasts but is most common in the Cape. *Call.* A loud 'ki-ok' and a short, repeated 'pok' alarm call. *Afrikaans.* Swartrugmeeu.

322 Caspian Tern *Hydroprogne caspia* L 50 cm
By far the largest tern of the region, this species has a black cap and a red bill which is usually black-tipped. The red of the bill has no hint of orange, a factor distinguishing this species from the smaller Royal Tern which also has a less massive bill and a more deeply forked tail. In flight, the tip of the underwing is black. The imm. has brown fringes to the wing coverts. *Habitat.* Large rivers, estuaries, bays, lagoons and inshore waters. Common on the east coast becoming less so on the west. Individuals occasionally venture to inland water bodies. *Call.* A hard, grating 'kraak'. The imm. whistles. *Afrikaans.* Reuse Sterretjie.

343 African Skimmer *Rynchops flavirostris* L 38 cm
The unique, peculiarly shaped red bill is diagnostic, the lower mandible being longer than the upper. At rest the black upperparts contrast with the white underparts and red bill. Imm. is like the ad. except that the black upperparts are replaced by pale-fringed brown feathering, and the bill is blackish, becoming brighter with age. *Habitat.* Large rivers, bays and lakes where there is suitable feeding and roosting habitat. Resident only in the very north of the region. *Call.* A harsh 'rak-rak'. *Afrikaans.* Waterploeër.

322▼ 312▲ 313▼

343▼

319 Blackheaded Gull *Larus ridibundus* L 40 cm

The ad. has a smaller body and bill than Greyheaded Gull and is paler grey above with a dark brown hood in breeding plumage. In non-breeding plumage is distinguished by the large wedge of white on the outer wing and the whitish, not grey underwing. In comparison with the ad., imm. has paler pink legs and bill base; less black on the upperwing and a much paler underwing. Franklin's Gull is always much darker above with black, not dark brown on the head. *Habitat.* Coasts and any body of inland water. A vagrant recorded in the Transvaal, Moçambique and Zimbabwe. *Call.* Vocal, with a typical small gull 'kraah'. *Afrikaans.* Swartkopmeeu.

316 Hartlaub's Gull *Larus hartlaubii* L 38 cm

In comparison with breeding Greyheaded Gull, is slightly smaller, has a thinner, duller bill, only a suggestion of a grey hood, dark, not silver eyes, and deeper red legs. In non-breeding plumage has a plain white head. Imm., compared with imm. Greyheaded Gull, has darker legs, lacks the two-tone bill, has less black on the tail and faint dark smudges on the head. Both Blackheaded and Franklin's gulls lack the dusky underwing. *Habitat.* Forages on the coasts and islands of the western Cape and Namibia. Vagrant to the east coast. *Call.* A typical 'karrh' and 'pok-pok'. *Afrikaans.* Hartlaubse Meeu.

318 Sabine's Gull *Larus sabini* L 34 cm

A small gull with buoyant, tern-like flight. The boldly tri-coloured wing and forked tail are diagnostic. Imm. has brownish back extending on to the inner forewing. Imm. resembles imm. Blacklegged Kittiwake but is smaller, has a darker head and upperwing, and lacks the blackish collar. In breeding plumage shows a dark grey hood and a yellow-tipped bill. *Habitat.* A totally pelagic gull. Common summer visitor to the Cape and Namibian coasts, rarer off the east coast. *Call.* Silent in our region. *Afrikaans.* Mikstertmeeu.

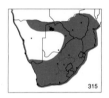

315 Greyheaded Gull *Larus cirrocephalus* L 42 cm

Breeding ad. differs from Hartlaub's Gull by having a much denser grey hood, a brighter red bill and legs, and silver, not dark eyes. It has a slightly larger body, head and bill than Hartlaub's Gull. Imm. Greyheaded has much darker, more extensive smudges on the head, paler legs, a pink bill with a dark tip, and more black on the tip of the tail than imm. Hartlaub's. Both Blackheaded and Franklin's gulls are smaller, have very different wing patterns and lack the dark underwing of the Greyheaded Gull. *Habitat.* Coastal and freshwater areas, and wetlands, in the east and along the Namibian coast. Least common in the Cape. *Call.* A typical 'karrh' and 'pok-pok'. *Afrikaans.* Gryskopmeeu.

320 Blacklegged Kittiwake *Larus tridactyla* L 40 cm

Easily confused with Sabine's Gull but is larger and ad. has a diagnostic all-yellow bill; squarish, not forked tail; and has black only on the wing tips. Imm. distinguished from imm. Sabine's by its larger size, paler head and upperwing with distinct black open 'M' pattern. *Habitat.* Coastal and pelagic. Vagrant to the western Cape. *Call.* 'Kitt-e-wake'. *Afrikaans.* Swartpootbrandervoël.

317 Franklin's Gull *Larus pipixcan* L 35 cm

Smaller and darker than Greyheaded and Hartlaub's gulls. Always displays at least a partial black hood and a whitish, not grey underwing. In breeding plumage has a full black hood, white eye-rings and a rosy hue to the breast. Much darker abǫve than Blackheaded, which has a brown, not black hood. The imm. Blackheaded has irregular dark head smudges, not the partial hood of the imm. of this species. In flight the wing pattern of the ad. is diagnostic: the black wing tip is separated from the grey inner wing by a white band, whereas the Blackheaded Gull has a large white wedge on the outer wing. *Habitat.* At sea or along the coast. Vagrant to the Cape and Natal. *Call.* Not recorded in our region. *Afrikaans.* Franklinse Meeu.

320 ▲   318 ▼

319 ▼   316 ▼

317 ▼   315 ▼

321 Gullbilled Tern *Gelochelidon nilotica* L 39 cm

A very pale, large, relatively long-legged tern with a short, stubby bill much like a gull's. In our region most often seen without the breeding plumage of black cap: instead, it shows a variable black smudge behind the eye. Sandwich Tern has a longer, yellow-tipped bill, and a deeply forked white tail, as opposed to the notched grey tail of the Gullbilled Tern. *Habitat.* Variable, ranging from fields, marshes and vleis to coasts. Uncommon in Botswana and a vagrant to Mocambique and Natal. *Call.* Variations on 'kek-kek'. *Afrikaans.* Oostelike Sterretjie.

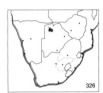

326 Sandwich Tern *Sterna sandvicensis* L 40 cm

A relatively large, very pale tern. The black, yellow-tipped bill is diagnostic. Could be confused with Gullbilled Tern which has a short, stubby bill, but Sandwich Tern shows a white, not grey rump, and a tail which is far more forked. In breeding plumage the breast may have a faint rosy hue. *Habitat.* Inshore waters, estuaries and bays. A common summer migrant to all coasts. *Call.* 'Kirik'. *Afrikaans.* Grootsterretjie.

325 Lesser Crested Tern *Sterna bengalensis* L 38 cm

Smaller and more graceful than Swift Tern from which it also differs by being paler above and by having a more slender orange, not yellow bill. Royal Tern is larger, even paler above and has a larger, deeper orange bill. Like the Swift Tern, the imm. has brown on the wing coverts, but it can be distinguished by the orange bill. *Habitat.* Bays, estuaries and inshore waters. This is a common summer migrant to Moçambique and Natal but is a vagrant to the Cape and is unrecorded in Namibia. *Call.* A hoarse 'kreck'. *Afrikaans.* Kuifkopsterretjie.

324 Swift Tern (Greater Crested Tern) *Sterna bergii* L 46 cm

Intermediate in size between Caspian and Sandwich terns. Has a large yellow, not orange bill, which distinguishes it from the Lesser Crested Tern which is also smaller and paler above. Royal Tern has much paler upperparts and a decidedly orange bill. In breeding plumage the Royal Tern has a black cap reaching the base of the bill, whereas the Swift Tern in breeding plumage shows a white frons. Imm. is barred dark brown and has a dusky yellow bill. *Habitat.* Inshore waters, the larger bays and estuaries. Breeds in the Cape and then disperses to all coasts. *Call.* Ad.'s call is a hard 'kee-eck'. Imm. gives a thin vibrating whistle. *Afrikaans.* Geelbeksterretjie.

331 Blacknaped Tern *Sterna sumatrana* L 30 cm

Intermediate in size between Little and Common terns. Ad. has a characteristic black band encircling the head from eye to eye, broadening on the nape. Upperparts are pale, resembling the Roseate Tern, but the latter has black streaking on the crown and is larger. *Habitat.* Coastal waters, sometimes roosting with other tern species in estuaries. A vagrant to Natal and Moçambique. *Call.* A clipped, repeated 'ki-ki'. *Afrikaans.* Swartkroonsterretjie.

323 Royal Tern *Sterna maxima* L 48 cm

A very pale tern with a deep orange bill, differing from the heavy, red or black-tipped bill of the larger Caspian Tern. In non-breeding plumage has a very extensive white forehead and crown never seen in the Caspian Tern. Most resembles the Lesser Crested Tern which is smaller, and darker grey above with a brighter, more slender orange bill. Compared with the Swift Tern, is paler above and has an orange, not yellow bill. *Habitat.* Coasts, bays and estuaries. A rare vagrant with one record from Swakopmund, Namibia. *Call.* A loud, harsh 'ree-ack'. *Afrikaans.* Koningsterretjie.

142

326 ▲   324 ▼ 325 ▼   321 ▲

323 ▼   331 ▼

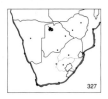

327 Common Tern *Sterna hirundo* L 33 cm
Differs from Arctic and Antarctic terns by its noticeably longer bill and legs and, in non-breeding plumage, further differs by its greyish, not white rump and tail. In comparison with Whitecheeked Tern, has a paler rump and tail which contrast with the back. In breeding plumage, differs from Arctic Tern by its darker outer wing and black-tipped, red bill. Roseate Tern has a longer, heavier bill, paler grey upperparts and pinkish underparts. *Habitat.* Open sea and coastal lakes. The most abundant tern on all coasts. Some birds overwinter. *Call.* 'Kik-kik' and 'ke-arh'. *Afrikaans.* Gewone Sterretjie.

328 Arctic Tern *Sterna paradisaea* L 33 cm
Obviously shorter legged and billed when compared with Common Tern, this species also shows a white, not pale grey rump and tail, and paler wing tips in flight. Most resembles Antarctic Tern but lacks the dark red on bill and legs, and is less thickset. Antarctic Tern attains breeding plumage from October to April and the Arctic from April to September, so confusion is unlikely. Whitecheeked Tern has a grey, not white rump and tail. Roseate Tern has a longer bill and legs, and lacks the dark tips to the outer primaries. *Habitat.* A pelagic tern which sometimes roosts ashore. Common summer migrant to all coasts. *Call.* A short 'kik-kik' in flight. *Afrikaans.* Arktiese Sterretjie.

329 Antarctic Tern *Sterna vittata* L 34 cm
More thickset than the Arctic and Common terns with a heavier, dusky red bill. Whitecheeked Tern has a grey, not white rump, and the Roseate Tern is paler and much more slender. Imm. has barred upperparts. *Habitat.* A winter visitor to coastal waters, it commonly roosts ashore in the Cape. Rare in Natal. *Call.* A sharp, high-pitched 'kik-kik'. *Afrikaans.* Grysborssterretjie.

330 Roseate Tern *Sterna dougallii* L 36 cm
Distinguished by pale upperparts, and long bill and tail. Blacknaped Tern is smaller and lacks the black on the crown. In breeding plumage it has long tail streamers, a pink flush to the breast, and a black-tipped red bill. Imm. differs from imm. Antarctic Tern by blacker cap, longer bill and more slender body. *Habitat.* Breeds on islands off Algoa Bay and is regular but uncommon in Natal. *Call.* A grating 'aarh'. *Afrikaans.* Rooiborssterretjie.

335 Little Tern *Sterna albifrons* L 23 cm
Differs from Damara Tern by its shorter, straight bill and slightly darker upperparts. The legs are brownish yellow and the bill frequently has a yellow base; legs and bill become yellower around March. *Habitat.* A summer visitor to the surf-line, large bays and estuaries. Common on the east coast, rare on the west. *Call.* A rasping 'kek-kek'. *Afrikaans.* Kleinsterretjie.

334 Damara Tern *Sterna balaenarum* L 23 cm
Differs from Little Tern by longer, slightly decurved bill and paler upperparts. In breeding plumage, shows black bill and legs, and a complete black cap. Imm. has brown barring on the upperparts: Little Tern does not occur in this plumage in our region. *Habitat.* Coasts, bays and lagoons from the eastern Cape to Namibia. Endemic but never forms large flocks. *Call.* A harsh 'kid-ick'. *Afrikaans.* Damarasterretjie.

336 Whitecheeked Tern *Sterna repressa* L 32 cm
The uniform colour of the back, rump and tail appears a 'dirtier' grey than the Common and Arctic terns, which have white tails and rumps. In breeding plumage resembles Antarctic Tern which has a white rump, and Whiskered Tern, from which it differs by its larger size, deeply forked tail and lack of white vent. *Habitat.* Inshore waters and estuaries. Rare vagrant: two records from Natal. *Call.* A ringing 'kee-leck'. *Afrikaans.* Witwangsterretjie.

327 ▲
329 ▼
336 ▼
328 ▲
335 ▼
330 ▼
334 ▼

333 Bridled Tern *Sterna anaethetus* L 35 cm

Similar to the larger Sooty Tern but has paler, brown-grey upperparts. The white forehead extends as an eyebrow stripe slightly behind the eye (Sooty Tern's forehead patch ends at the eye). Imm. has the wing coverts finely edged buffy and its underparts are white, whereas imm. Sooty Tern's are partially black. *Habitat.* Pelagic, occasionally roosts ashore. Vagrant to the Natal coast and the Cape. *Call.* 'Wup-wup'. *Afrikaans.* Brilsterretjie.

332 Sooty Tern *Sterna fuscata* L 44 cm

May be confused with Bridled Tern but is larger, the white forehead extends only as far as the eye and the black crown does not contrast with the dark brown back. Imm. has black underparts whereas Bridled Tern is always white below. *Habitat.* Pelagic but adverse weather brings it ashore. A vagrant to the coast and inland to the Transvaal and Zimbabwe. *Call.* Variations on 'wick-a-wick'. *Afrikaans.* Roetsterretjie.

341 Lesser Noddy *Anous tenuirostris* L 32 cm

Smaller than the Common Noddy with a longer, thinner bill. The whitish forehead merges with the brown lores. The underwing is dark brown, not two-toned as in the Common Noddy. Imm. is distinguished by smaller size, longer bill, dark underwing and more extensive, pale forehead. *Habitat.* Pelagic, but occasionally roosts ashore. A vagrant to Natal and Moçambique. *Call.* Not recorded in our region. *Afrikaans.* Kleinbruinsterretjie.

340 Common Noddy (Brown Noddy) *Anous stolidus* L 42 cm

A brown, tern-like bird with a wedge-shaped tail, notched only when fanned. It could be confused with the Lesser Noddy which is smaller and appears more uniformly dark in flight. The white forehead contrasts sharply with the brown lores. In flight the underwing is pale with dark margins. The imm. differs from imm. Lesser Noddy by its larger size, stouter bill and darker forehead. *Habitat.* Pelagic but occasionally roosts ashore. A rare vagrant to Natal and the Cape. *Call.* A hoarse 'kark'. *Afrikaans.* Grootbruinsterretjie.

339 Whitewinged Tern *Chlidonias leucopterus* L 23 cm

When breeding resembles Black Tern but has a pied underwing and paler back and wings. In non-breeding plumage resembles Whiskered Tern but has a white, not grey rump. Non-breeding Black Tern is much darker, having more black on the head and a black shoulder smudge. *Habitat.* A widely distributed summer migrant. *Call.* A short 'kek-kek'. *Afrikaans.* Witvlerksterretjie.

337 Black Tern *Chlidonias niger* L 22 cm

In breeding plumage differs from Whitewinged Tern by uniform, not pied underwing, and upperwings which do not contrast with the back. In non-breeding plumage very similar to Whitewinged Tern but the dark shoulder smudge is diagnostic, it shows more black on the head, and there is no contrast between the back, rump and tail. Whiskered Tern is paler, has less black on the head and no shoulder smudge. *Habitat.* Common off Namibian coast in summer. *Call.* Silent in our region. *Afrikaans.* Swartsterretjie.

338 Whiskered Tern *Chlidonias hybridus* L 25 cm

In breeding plumage resembles Whitecheeked Tern but is smaller, has a white vent and a notched tail. In non-breeding plumage very similar to Whitewinged Tern but has a duller white rump. Black Tern is darker with more black on the head and has a dark shoulder smudge. Imm. similar to imm. Antarctic Tern but the latter has a white rump. *Habitat.* Common on suitable bodies of inland water. *Call.* A repeated, hard 'zizz'. *Afrikaans.* Witbaardsterretjie.

342 Fairy Tern (White Tern) *Gygis alba* □

No positive records from our region. *Afrikaans.* Feesterretjie.

333▲ 332▼ 339▲

341▼ 337▼

340▼ 338▼

344 Namaqua Sandgrouse *Pterocles namaqua* L 25 cm
In the region, the only sandgrouse which has a long pointed tail. At rest or when walking it most resembles the Doublebanded Sandgrouse but lacks the white markings on the head and the black and white breast band. Female and imm. differ from female and imm. Doublebanded Sandgrouse by being more buffy yellow on throat and breast and by their pointed, not rounded tails. *Habitat.* Grasslands, and true and semi-desert. Avoids mountainous and wet regions. Common endemic to the dry west, absent from the east and north-east. *Call.* A nasal 'kalke-ven, kalke-ven' flight call. *Afrikaans.* Kelkiewyn.

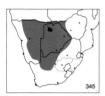

345 Burchell's Sandgrouse (Spotted Sandgrouse) *Pterocles burchelli* L 25 cm
The white-spotted cinnamon breast and belly, combined with the white-spotted back and wing coverts, render this small sandgrouse unmistakable. Female and imm. resemble ad. male but lack the blue-grey throat and are generally drabber in coloration. *Habitat.* Lightly wooded, dry areas and grass-covered sand dunes. Uncommon endemic to the dry central and northern regions. *Call.* A soft, mellow 'chup-chup, choop-choop' given in flight and around waterholes. *Afrikaans.* Gevlekte Sandpatrys.

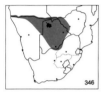

346 Yellowthroated Sandgrouse *Pterocles gutturalis* L 30 cm
The largest sandgrouse of the region and easily identified in flight by its very dark belly and almost black underwings. The creamy yellow throat is bordered by a thin black band. Female and imm. also show the very dark belly and underwing, which eliminates confusion with any other imm. or female sandgrouse. *Habitat.* Grassy, sandy areas and light, dry broad-leafed woodland. Uncommon in the north and north-western regions. *Call.* A 'tweet-weet, tweet-weet' flight call has been described. *Afrikaans.* Geelkeelsandpatrys.

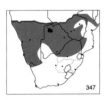

347 Doublebanded Sandgrouse *Pterocles bicinctus* L 25 cm
Most resembles the Namaqua Sandgrouse but the male is easily identified by the bold black and white markings on the head and by a thin black and white breast band. Female and imm. distinguished from female and imm. Namaqua Sandgrouse by having a darker streaked crown; barred, not streaked breast; and round, not pointed tail. *Habitat.* Dry, lightly wooded areas and thornveld. Common in the northern part of the region. *Call.* A whistling 'chwee-chee-chee' and a soft 'wee-chee-choo-chip-chip' flight call. *Afrikaans.* Dubbelbandsandpatrys.

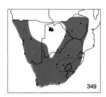

349 Rock Pigeon (Speckled Pigeon) *Columba guinea* L 33 cm
The reddish wings spotted with white, and the bare red patches around the eyes are diagnostic. The only similar pigeon in the region is the Rameron which is much larger, darker and has bright yellow bare eye patches. *Habitat.* Mountain ranges, rocky terrain, coastal cliffs and cities. Common in suitable habitat but absent from the eastern and central areas. *Call.* A deep, booming 'hooo-hooo-hooo' and a softer 'coocoo-coocoo'. *Afrikaans.* Kransduif.

350 Rameron Pigeon *Columba arquatrix* L 42 cm
The largest pigeon of the region. Easily identified by its dark-coloured plumage, finely speckled with white, conspicuous yellow patch surrounding the eyes, and yellow bill, legs and feet. In flight it appears as a large, dark blue or black pigeon. Imm. resembles ad. but has duller coloured eye patches, bill, legs and feet. *Habitat.* Evergreen forests and exotic plantations, especially where stands of Bugweed *Solanum mauritianum* occur. Common in suitable habitat in the south and east of the region. *Call.* A low but raucous 'coo'. *Afrikaans.* Geelbekbosduif.

148

347▲   344▼ 345▼   346▲

349▼ 350▼

351 Delegorgue's Pigeon (Bronzenaped Pigeon) *Columba delegorguei*
L 30 cm
In the field appears as dark as the Rameron Pigeon but is much smaller and the male has a diagnostic, pale half-moon patch on the hind collar. The iridescent patches on the sides of the neck are visible only at close range. Female and imm. lack the pale hind collar and differ from the similar-sized Cinnamon Dove by having a much darker head and body. *Habitat.* Evergreen coastal and inland forests. Uncommon in Natal and Zululand, with scattered records in the eastern Transvaal and Zimbabwe. *Call.* A low, soft, frequently repeated 'duu-duu-duu' in the early morning. *Afrikaans.* Withalsbosduif.

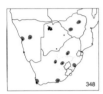

348 Feral Pigeon *Columba livia* L 33 cm
Feral pigeons are wild descendants of the domesticated Rock Doves of Europe. Plumage very variable, with black, blue, grey, white and reddish forms occurring. The blue form is identical to the true Rock Dove: bluish grey with black bars on wings and tail, white rump patch, and glossy green and purple on sides of neck. *Habitat.* Breeds in the wild on coastal cliffs in Transkei. Abundant throughout the region in cities, towns and some villages. *Call.* Typical domestic pigeon 'coo-roo-coo'. *Afrikaans.* Tuinduif.

353 Mourning Dove *Streptopelia decipiens* L 30 cm
Distinguished from the similar Cape Turtle Dove by having red skin around yellow eyes and by having a totally grey head. Confusion might arise with the Redeyed Dove but that species is much larger, is overall very much darker and has a deep red, not pale-coloured eye. *Habitat.* Thornveld, riverine forest, and cultivated areas and gardens in bushveld. Locally common in the east and north. Particularly common at Satara Camp in the Kruger National Park. *Call.* A soft, dove-like 'coooc-currr'. *Afrikaans.* Rooioogtortelduif.

352 Redeyed Dove *Streptopelia semitorquata* L 35 cm
Much larger and darker than the similar Cape Turtle Dove. In flight, although both species have a grey band on the tip of the undertail, the Redeyed can be distinguished by the lack of any white on the uppertail. At close range, the red eye and red skin around the eye, are diagnostic. It is much larger, and darker in colour than the Mourning Dove. *Habitat.* Found in a range of areas from dry bushveld to coastal forests, and has adapted to city gardens and open parks. Common in the south, east and north of the region, being absent from the drier west. *Call.* Alarm call is a 'chwaa'. Other calls are variable but a very dove-like 'coo-coo, kook-co-co' is typical. *Afrikaans.* Grootringduif.

354 Cape Turtle Dove *Streptopelia capicola* L 28 cm
The white-tipped tail, which is conspicuous in flight, is diagnostic. Much smaller and paler than the Redeyed Dove which lacks the white tail tip. Distinguished from the Mourning Dove by having a paler grey head and by lacking red skin around the eyes. *Habitat.* Found in virtually every habitat in the region but avoids dense coastal forests. Abundant throughout. *Call.* A harsh 'kurrrr' when alarmed, and the well-known dove-call of Africa: 'kuk-cooo-kuk'. *Afrikaans.* Gewone Tortelduif.

361 Green Pigeon *Treron calva* L 30 cm
When seen clambering through the forest canopy, the immediate impression is of a green, parrot-like bird. At closer range this species is unmistakable with its green and yellow plumage and chestnut vent. When feeding, it climbs around fruiting trees, sometimes hanging upside-down on branches to glean fruit. *Habitat.* Evergreen, coastal and thick riverine forests. Common in the north and east. *Call.* A series of liquid whistles 'thweeeloo, tleeeoo'. *Afrikaans.* Papegaaiduif.

150

348 ▲   353 ▼   352 ▼   351 ▲

354 ▼   361 ▼

355 Laughing Dove (Palm Dove) *Streptopelia senegalensis* L 26 cm
Distinguished from the larger Cape Turtle Dove by the lack of black hind collar and by having a diagnostic black-speckled necklace across its cinnamon breast, and by its cinnamon-coloured back. In flight the obvious blue-grey forewings are conspicuous and the white tip and sides of the tail are shown. *Habitat.* Found in a wide range of habitats but avoids true desert. The most common and best-known dove in the region, having adapted to gardens and city centres throughout. *Call.* Common call is a distinctive rising and falling 'ooo-coooc-coooc-coo-coo'. *Afrikaans.* Rooiborsduifie (Lemoenduifie).

357 Bluespotted Dove *Turtur afer* L 22 cm
Difficult to differentiate from Greenspotted Dove except at close range when the yellow-tipped red bill and blue wing spots are seen. In flight its back and rump appear more rufous than those of the Greenspotted Dove. Imm. lacks the bill colouring and blue wing spots of ad. and is virtually indistinguishable from imm. Greenspotted. *Habitat.* Moist broad-leafed woodlands and along rivercourses in evergreen forests. Uncommon and found only in the extreme north-east. *Call.* A series of muffled 'du-du-du-du' call notes, similar to those of Greenspotted Dove. *Afrikaans.* Blouvlekduifie.

358 Greenspotted Dove (Emeraldspotted Dove) *Turtur chalcospilos* L 22 cm
At close range the dark-coloured bill and green wing spots distinguish this species from the Bluespotted Dove. Normally avoids the evergreen forests frequented by the Bluespotted Dove. *Habitat.* Thornveld, dry broad-leafed woodland, and riverine forest situated in these habitats. Common in suitable habitat in the eastern, south-eastern, central and northern parts of the region. *Call.* One of the most characteristic calls of the bushveld: a series of low 'du-du-du-du' notes which descend in scale and quicken towards the end. *Afrikaans.* Groenvlekduifie.

359 Tambourine Dove *Turtur tympanistria* L 22 cm
The white face and underparts are diagnostic. In flight the chestnut underwings contrast strongly with the white belly. In comparison with the male, the female and imm. are slightly darker below but still have paler faces and underparts than all other small doves. *Habitat.* Dense evergreen, riverine and coastal forests. Common in suitable habitat in the south-eastern, eastern and north-eastern parts of the region. *Call.* A series of 'du-du-du' notes similar to those of the Greenspotted Dove but, instead of descending in scale, this call trails off at the end. *Afrikaans.* Witborsduifie.

356 Namaqua Dove *Oena capensis* L 28 cm
The smallest-bodied dove of the region and the only one to have a long pointed tail, and a black face and throat. In flight the long tail, combined with white underparts and chestnut flight feathers, render this bird unmistakable. Female and imm. lack the black face of the male and have slightly shorter tails. *Habitat.* Prefers drier regions such as thornveld, scrub and true desert. Found throughout the region, it is common in the south and west, but subject to local movements. *Call.* A soft, low 'coooo-hoooo'. *Afrikaans.* Namakwaduifie.

360 Cinnamon Dove *Aplopelia larvata* L 26 cm
When flushed from the forest floor, it 'explodes' and dashes off swiftly, giving the impression of a small dark dove which lacks the pale, barred rump of the Blue- and Greenspotted doves. When glimpsed at rest on the forest floor, the whitish face and cinnamon underparts are diagnostic. *Habitat.* Inland evergreen and coastal forests. Common in suitable habitat in the extreme south, east and north-east of the region. *Call.* A low, somewhat raspy 'hooo-oooo' is given from dense thickets. *Afrikaans.* Kaneelduifie.

152

357 ▲   358 ▲   359 ▲

356 ▲   360 ▼   355 ▼

364 Meyer's Parrot *Poicephalus meyeri* L 22 cm

May be confused with Rüppell's Parrot in Namibia, where their ranges overlap. Differs from Rüppell's Parrot by having a green, not blue rump and belly, and a brown, not grey head, with a yellow bar across the crown. Imm. resembles ad. but is duller in coloration and lacks the yellow bar across the crown. *Habitat.* Occurs in dry broad-leafed woodland and flocks regularly congregate at waterholes. Common in certain localities in northern and central regions. *Call.* A loud, piercing 'chee-chee-chee-chee' and various other screeches and squawks. *Afrikaans.* Bosveldpapegaai.

365 Rüppell's Parrot *Poicephalus rueppellii* L 22 cm

At rest, best distinguished from Meyer's Parrot by greyish, not brown throat and head, and by its blue, not green belly. In flight easily differentiated from Meyer's Parrot by obvious blue, not green rump. The yellow bar across the crown of Meyer's Parrot is absent in this species. Imm. resembles ad. *Habitat.* Dry woodland, thornveld and dry rivercourses and, in the north, shows a preference for stands of baobab trees. Uncommon endemic, thinly distributed in northern Namibia. *Call.* Screeches and squawks similar to those of Meyer's Parrot. *Afrikaans.* Bloupenspapegaai.

367 Rosyfaced Lovebird *Agapornis roseicollis* L 18 cm

Usually located by their screeching calls as they are very difficult to detect when they sit motionless in a leafy tree or bush. Flight is extremely rapid, and the blue rump shows up clearly against the green back. There is no range overlap with the similar Lilian's Lovebird which has a green, not blue rump. *Habitat.* Dry broad-leafed woodland, semi-desert and mountainous terrain. Common endemic to the north-west. *Call.* Typical parrot-like screeches and shrieks. *Afrikaans.* Rooiwangparkiet.

369 Blackcheeked Lovebird *Agapornis nigrigenis* L 14 cm

The dark brown head contrasting with a bright red bill render this small lovebird unmistakable. Imm. has a dark head like the ad. but its bill is dark grey, not red. *Habitat.* Riverine forests and open woodland. Rare and very local in the Caprivi Strip region and near Victoria Falls. *Call.* Shrieking identical to that of Lilian's Lovebird. *Afrikaans.* Swartwangparkiet.

368 Lilian's Lovebird *Agapornis lilianae* L 14 cm

Slightly smaller than similar Rosyfaced Lovebird, this species differs by having a green, not blue rump. Unlikely to be found with Rosyfaced Lovebird as their distributions do not overlap. *Habitat.* Open broad-leafed woodland and thornveld. Occurs rarely along the Zambezi valley in Zimbabwe and Moçambique. *Call.* A high-pitched, staccato shrieking. *Afrikaans.* Niassaparkiet.

366 Roseringed Parakeet *Psittacula krameri* L 40 cm

The only parakeet in the region to have an extremely long, pointed tail. At close range the dark red bill and dark ring around the neck are diagnostic of this bright green species. *Habitat.* Small numbers in coastal forest but mostly in urban and suburban parks and gardens. Uncommon and local with a small population established in Durban and at Sodwana Bay, Natal. *Call.* Various shrieks and screams, being particularly vocal at roost. *Afrikaans.* Ringnekpapegaai.

364 ▼

◀366

365 ▼

367 ▼

369 ▼

368 ▼

363 Brownheaded Parrot *Poicephalus cryptoxanthus* L 24 cm
The greenest parrot in our region, this species has a uniform brown head, and bright yellow underwings visible in flight. Imm. resembles ad. but is generally duller with less vivid yellow underwings. *Habitat.* Thornveld, riverine forest and open woodland. Common in the east and north-east. *Call.* A typically parrot-like raucous shriek. *Afrikaans.* Bruinkoppapegaai.

362 Cape Parrot (Brownnecked Parrot) *Poicephalus robustus* L 35 cm
The largest parrot in the region. The combination of its size, red shoulders, forehead and throat is diagnostic. Imm. lacks the red shoulders and forehead but its size and massive bill should rule out confusion with other parrots in the region. *Habitat.* Evergreen forests and thick woodland but has adapted to exotic plantations in the east. Uncommon and thinly distributed in the east and north-east. *Call.* Various loud, harsh screeches and squawks. *Afrikaans.* Grootpapegaai (Woudpapegaai).

370 Knysna Lourie *Tauraco corythaix* L 46 cm
The all-green head, white eye-ring and white tips to the head crest distinguish this species from the Purplecrested Lourie. The race on the east coast, north of St Lucia, has an elongated head crest. In flight shows conspicuous crimson patches on primaries. *Habitat.* Dense evergreen forests. Common in the eastern, north-eastern and northern parts of the region. *Call.* A loud, booming 'kow-kow-kow-kow' and a quieter 'krrr' alarm note. *Afrikaans.* Knysnaloerie.

371 Purplecrested Lourie *Tauraco porphyreolophus* L 46 cm
Appears generally darker than the very green-looking Knysna Lourie. Differs from that species mainly by its purple head crest, which appears black unless seen in good light, and by the lack of white around the eyes and on the crest. Like the Knysna Lourie, it is very furtive and is usually seen only when leaping from tree to tree, showing the very conspicuous red in its wings.
Habitat. Evergreen coastal, inland and riverine forests. Common in the east, north-east and north of the region. *Call.* A loud 'kok-kok-kok-kok'. *Afrikaans.* Bloukuifloerie.

373 Grey Lourie *Corythaixoides concolor* L 48 cm
This ash-grey bird with its long tail and crest is one of the more obvious birds of the bushveld. Vocal and conspicuous, they are often seen in small groups perched high in thorn trees. *Habitat.* Thornveld and dry, open woodland. Common in the northern sector of the region, absent from the south.
Call. A harsh, nasal 'waaaay' or 'kay-waaaay', from which it derives its vernacular name of the 'go-away' bird. *Afrikaans.* Kwêvoël.

372 Ross's Lourie *Musophaga rossae* L 50 cm
The only dark blue lourie of the region and unlikely to be confused with any other lourie. The combination of yellow face and red crest is diagnostic. *Habitat.* Riverine forests. Vagrant to northern Botswana. *Call.* Described as a loud cackling. *Afrikaans.* Rooikuifloerie.

362▲ 370▼ 373▼ 363▲

371▼ 372▼

374 European Cuckoo *Cuculus canorus* L 33 cm
In the field virtually indistinguishable from the African Cuckoo except at very close range when the black bill with only a small amount of yellow at the base, is noticeable. The 'grey' cuckoos which are sometimes common in thornveld during summer, are this species. *Habitat.* Bushveld and dry woodland. Common summer migrant, absent from the central and western areas. *Call.* Silent in our region. *Afrikaans.* Europese Koekoek.

375 African Cuckoo *Cuculus gularis* L 33 cm
The call of this species is its most diagnostic feature. At close range it can be seen that the bill differs from that of the very similar European Cuckoo by being almost wholly yellow with a black tip. The female of both the European Cuckoo and this species have a rare rufous form. *Habitat.* Bushveld, open woodland and exotic plantations. Uncommon summer visitor to the north and north-east. *Call.* Similar to the Hoopoe's 'hoop-hoop' call. Female utters a fast 'kik-kik-kik'. *Afrikaans.* Afrikaanse Koekoek.

376 Lesser Cuckoo *Cuculus poliocephalus* L 28 cm
Not readily distinguishable from the African or European cuckoos except under optimum viewing conditions when its smaller size, and paler grey back, nape and crown become apparent. In flight the darker tail and rump contrast with the paler back, a feature not evident in either the African or European cuckoos. *Habitat.* Dense evergreen forests. A rare vagrant from Madagascar or Asia, with summer and winter sightings. *Call.* Silent in our region. *Afrikaans.* Kleinkoekoek.

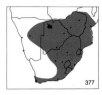

377 Redchested Cuckoo *Cuculus solitarius* L 30 cm
The characteristic trisyllabic call during summer indicates this cuckoo's presence, otherwise it is usually very difficult to locate as it sits motionless in thick canopy foliage. When seen, the chestnut breast is diagnostic. *Habitat.* Evergreen forests and has adapted well to exotic plantations. Common summer visitor but absent from the drier west and treeless regions. *Call.* A loud, often repeated 'weet-weet-weeoo'. *Afrikaans.* Piet-my-vrou.

378 Black Cuckoo *Cuculus clamosus* L 30 cm
The only all-black cuckoo in the region. Similar to dark phase Jacobin Cuckoo but differs by lacking the crest and white patches in the wings. Some ads. show rufous barring on the underparts. Imm. is brown, and lacks the indistinct white tail spots of ad. *Habitat.* Woodland and forest habitats, exotic plantations and suburban gardens. Common summer visitor throughout, except to the dry south and west. *Call.* A droning 'whoo-wheee-whoo-whoo-whee', or a fast 'yow-yow-yow-yow'. *Afrikaans.* Swartkoekoek.

382 Jacobin Cuckoo *Clamator jacobinus* L 34 cm
Dark phase birds differ from the similar Black Cuckoo by the noticeable crest and the white patches on the primaries. Pale phase birds resemble Striped Cuckoo but are pure white below, lacking any striping on throat and breast. *Habitat.* Found in woodlands, preferring thornveld. Common summer visitor throughout except to the dry centre and west. *Call.* A frequently repeated 'klee-klee-kleeuu-kleeuu'. *Afrikaans.* Bontnuwejaarsvoël.

381 Striped Cuckoo *Clamator levaillantii* L 38 cm
Unlikely to be confused with any other cuckoo except the pale phase Jacobin from which it differs by having diagnostic heavy black striping on the throat and breast. Imm. is browner above, very buff below but still shows the diagnostic throat striping. *Habitat.* Thornveld, riverine forest and open, broad-leafed woodland. Uncommon summer visitor and thinly distributed in the north, east and north-east. *Call.* A loud 'kleeo-kleeo-kleeo' and a faster 'che-che-che-che'. *Afrikaans.* Gestreepte Nuwejaarsvoël.

374 ▲   375 ▼

382 ▲   378 ▼

376 ▼

381 ▼

377 ▼

380 Great Spotted Cuckoo *Clamator glandarius* L 39 cm

This large cuckoo is unmistakable with its white-spotted dark back, elongated, wedge-shaped tail and long grey crest. Imm. is also heavily spotted on the back but has a small black crest, buffish underparts and rufous patches on the primaries. *Habitat.* Open woodland, forest edges and exotic plantations. Common summer visitor throughout, but mostly absent from the extreme south and the dry west. *Call.* A loud, far-carrying 'keeow-keeow-keeow' and a shorter, crow-like 'kark'. *Afrikaans.* Gevlekte Koekoek.

383 Thickbilled Cuckoo *Pachycoccyx audeberti* L 34 cm

Resembles the Great Spotted Cuckoo but differs by having a uniform grey, not white-spotted back and wing coverts, and by lacking a crest. Has a noticeably thick and heavy bill. The imm. is very striking with its white head flecked with black, and white-spotted upperparts: it differs from the imm. Great Spotted Cuckoo by lacking the rufous wing patches and the black cap.
Habitat. Forests and open woodland. Rare summer visitor to the north-east.
Call. A repeated 'chee-cher-cher' and a 'wee-yes-yes'.
Afrikaans. Dikbekkoekoek.

379 Barred Cuckoo *Cercococcyx montanus* L 33 cm

Much smaller-bodied than either the African or European Cuckoo, with an exceptionally long, narrow tail. Often located by its call but, when glimpsed, the brownish, heavily barred upperparts, and the broadly barred underparts, combined with the long tail and slender body, are diagnostic.
Habitat. Evergreen forests and open, broad-leafed woodland. Status uncertain: rare resident or summer visitor to eastern Zimbabwe and Moçambique just south of the Zambezi River. *Call.* A long series of 'cheee-phweew's, rising to a crescendo. Also a shorter 'hwee-hooa' or 'hwee-hooo'.
Afrikaans. Langstertkoekoek.

384 Emerald Cuckoo *Chrysococcyx cupreus* L 20 cm

Male is unmistakable with brilliant emerald green throat and breast and sulphur yellow belly. Female differs from female Klaas's and Diederik cuckoos by lacking white behind the eye and by having more brownish bronze upperparts and more heavily barred underparts. *Habitat.* Frequents the canopy of thick forests. Common summer visitor to the north, east and north-east. *Call.* A loud, ringing whistle 'wheet-huiee-wheet', often rendered as 'sweet georg-eee'. *Afrikaans.* Mooimeisie.

385 Klaas's Cuckoo *Chrysococcyx klaas* L 18 cm

Male differs from the very similar male Diederik Cuckoo by having only a small amount of white behind the eye, no white markings on the wings, and a brown, not red eye. Female differs from female Diederik Cuckoo by having finer, more extensive barring on flanks and chest, and by lacking white wing markings. *Habitat.* Forests, woodland, parks and gardens. Common in the north, east and south, but absent from the dry west. More obvious during summer when it calls. *Call.* A soft 'huee-jee' repeated five or six times. *Afrikaans.* Meitjie.

386 Diederik Cuckoo *Chrysococcyx caprius* L 18 cm

Male easily distinguished from similar Klaas's Cuckoo by the broad white eye-stripe, white spots on forewings and by its red, not brown eyes. Female differs from female Klaas's Cuckoo by being much greener above and whiter below, with much reduced barring on the flanks, and having white spots on the forewings. Imm. differentiated from imm. Klaas's Cuckoo by having a conspicuous red, not black bill. *Habitat.* Open grasslands with stands of trees, and thornveld and exotic plantations. Common summer visitor throughout the region. *Call.* A clear, persistent 'dee-dee-deedereek'. *Afrikaans.* Diederikkie.

383 ▲   380 ▼   384♀ ▼   379 ▲

385♀ ▼   386 ▼

388 **Black Coucal** *Centropus bengalensis* L 35 cm
A small coucal and the only one in the region to have a black head and body contrasting with rich chestnut back and wings. Non-breeding and imm. birds differ from other coucals by their small size, small bill and clear, buff-streaked upperparts. *Habitat.* Moist grasslands with thick stands of rank vegetation. Uncommon in the north-east. *Call.* A typical bubbling coucal call and other 'poopoop' and 'cuik' call notes. *Afrikaans.* Swartvleiloerie.

389 **Copperytailed Coucal** *Centropus cupreicaudus* L 48 cm
The largest coucal of the region, this species most resembles the Burchell's Coucal but it has a longer, broader and floppier tail which, like the head, is black with a coppery sheen. Imm. is the same size as ad. (thus ruling out confusion with other imm. coucals), and its long tail is barred black and brown. *Habitat.* Marshlands, thick reedbeds and adjoining bush. Common in the swampy regions of northern Botswana. *Call.* A louder, more resonant bubbling call than other coucals. *Afrikaans.* Grootvleiloerie.

390 **Senegal Coucal** *Centropus senegalensis* L 40 cm
Where the ranges of this species and Burchell's Coucal overlap, identification difficulties arise. This species differs only by its dark rump, and by having the base of its tail dark, not barred rufous as in Burchell's Coucal. Imms. of both species are indistinguishable in the field. Differs from Copperytailed Coucal by smaller size and brighter chestnut wings and back. *Habitat.* Tangled vegetation and long grass near water. Uncommon resident in the extreme north. *Call.* Bubbling call note, identical to that of Burchell's Coucal. *Afrikaans.* Senegalvleiloerie.

391 **Burchell's Coucal (Whitebrowed Coucal)** *Centropus superciliosus* L 44 cm
Dark phase birds are very similar to the slightly smaller Senegal Coucal but have fine rufous barring on the rump and base of tail. Pale phase birds have a diagnostic white eye-stripe and a white-flecked dark head and nape. In the field, imm. indistinguishable from imm. Senegal Coucal. *Habitat.* Long grass, riverine scrub and reedbeds. Common resident in the south, east and north, avoiding the drier areas. *Call.* A liquid bubbling 'doo-doo-doo-doo', descending in scale, then rising towards the end of the phrase. *Afrikaans.* Gewone Vleiloerie.

387 **Green Coucal** *Ceuthmochares aereus* L 33 cm
This small, shy bird is difficult to see in the thick, tangled undergrowth it frequents and is usually located by its call. When flushed, the dull green upperparts and long tail, combined with the yellow bill, are diagnostic. Imm. resembles ad. but has a duller bill. *Habitat.* Thick evergreen, riverine and coastal forests. Common in the lowlands of the north-east and east. *Call.* A clicking 'kik-kik-kik', winding up to a loud 'cher-cher-cher-cher'. *Afrikaans.* Groenvleiloerie.

162

389 ▲

391 ▼

387 ▲

388 ▼

390 ▼

392 **Barn Owl** *Tyto alba* L 34 cm
This golden buff and white owl could be confused with the Grass Owl but is much paler and has a less definite contrast between the upper- and underparts. The heart-shaped white facial disc shows the unusually small black eyes. *Habitat.* Often found near human habitation but also roosts in caves, hollow trees and mine shafts. Common throughout the region. *Call.* Many and varied calls, the most usual being an eerie 'shreeee'. *Afrikaans.* Nonnetjie-uil.

393 **Grass Owl** *Tyto capensis* L 36 cm
Although it could be confused with the Barn Owl, this species' much darker upperparts contrast more markedly with its white underparts. The Marsh Owl, found in the same habitat, has dark, not white underparts and noticeably rounded wings. *Habitat.* Grassy marshes and long grasslands but avoids thick stands of reedbeds. Uncommon resident found in the north-east, east and south. *Call.* Normally silent but hisses loudly when disturbed at nest. *Afrikaans.* Grasuil.

401 **Spotted Eagle Owl** *Bubo africanus* L 43-50 cm
Grey and rufous colour phases occur. Grey phase is the most common and is distinguished from the Cape Eagle Owl by its smaller size, lack of dark brown breast patches, by its finely barred belly and flanks, by its yellow, not orange eyes, and its smaller feet. Although the rufous phase closely resembles the Cape Eagle Owl, it is separable on the above characteristics except that the eyes are a similar orange. *Habitat.* Frequents a wide range of habitats but avoids thick forest. The most common 'eagle owl' of the region, it is found throughout. *Call.* Call similar to that of Cape Eagle Owl but is a shorter 'huu-whooo'. *Afrikaans.* Gevlekte Ooruil.

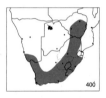

400 **Cape Eagle Owl** *Bubo capensis* L 48-54 cm
Not easily distinguished from the more common Spotted Eagle Owl unless either seen at close range or heard calling. The Cape Eagle Owl, and in particular the race **B.c. mackinderi** in Zimbabwe, is on average larger than the Spotted Eagle Owl. It has brown blotching on the breast forming two distinct dark areas; bold, not fine barring on belly and flanks; and much larger feet than the Spotted Eagle Owl. At close range the orange, not yellow eyes can be seen. *Habitat.* Rocky and mountainous terrain. Considered rare but may have been overlooked. Found in the south, east, and north to Zimbabwe. *Call.* A loud, far-carrying 'hu-hooooo'. *Afrikaans.* Kaapse Ooruil.

402 **Giant Eagle Owl** *Bubo lacteus* L 60-65 cm
The largest owl of the region, its large size and general grey coloration make it easily identifiable. The northern race of the Cape Eagle Owl approaches it in size but that species has rufous coloration and boldly barred underparts. *Habitat.* Dry, broad-leafed woodland, thornveld and riverine forest. Uncommon and thinly distributed throughout, except in the south. *Call.* A pig-like 'unnh-unnh-unnh' grunting. *Afrikaans.* Reuse Ooruil.

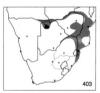

403 **Pel's Fishing Owl** *Scotopelia peli* L 63 cm
The large size and tawny rufous coloration render this species unmistakable. When alarmed, the bird fluffs up its head feathers giving it a huge, round-headed appearance. At rest, the large, dark brown eyes dominate the unmarked tawny facial disc. The unfeathered legs and feet are difficult to see in the field. *Habitat.* Large trees around lakes and slow-moving rivers. Uncommon and thinly distributed in the north and east. *Call.* A deep, booming 'hoo-huuuum' and a jackal-like wailing. *Afrikaans.* Visuil.

164

392▲ 400▼ 402▼ 393▲ 403▼ 401▲

396 **Scops Owl** *Otus senegalensis* L 20 cm
This small owl could be confused with the Whitefaced Owl because both have 'ear' tufts, but this species has a grey, not white face and is considerably smaller and slimmer. A grey phase and brown phase occur. *Habitat.* Bushveld and dry, open woodland. Absent from forested regions. Common in the north and east. *Call.* A soft, croaking, frog-like 'prrrup-prrrrup'. *Afrikaans.* Skopsuil.

397 **Whitefaced Owl** *Otus leucotis* L 28 cm
The only other small owl with 'ear' tufts is the Scops, from which this species differs by having a conspicuous white facial disc edged with black, and by its bright orange, not yellow eyes. It is also larger, and much paler grey than the Scops Owl. *Habitat.* Thornveld and dry broad-leafed woodland. Common throughout except in the south. *Call.* A fast, hooting 'doo-doo-doo-doo-hohoo' call. *Afrikaans.* Witwanguil.

399 **Barred Owl** *Glaucidium capense* L 21 cm
Might be confused with Pearlspotted Owl from which it differs by having barred upperparts and tail, and conspicuous white edging to the scapulars. Has a disproportionately large, rounded head, which should further help distinguish it from Pearlspotted Owl. *Habitat.* Bushveld and open woodland, especially the taller trees of dry river beds. Common in suitable habitat in the north and east, with a small population in the eastern Cape. *Call.* A soft, frequently repeated 'kerrr-kerrr-kerrr' and a 'trrru-trrre'.
Afrikaans. Gebande Uil.

398 **Pearlspotted Owl** *Glaucidium perlatum* L 18 cm
The smallest owl of the region. The rounded head with no 'ear' tufts, and the white spotting on back and tail distinguish this species from the Scops and Whitefaced owls. Differentiated from Barred Owl by its smaller size and lack of barring on the upperparts. Shows two black spots on the nape.
Habitat. Primarily a dry thornveld and broad-leafed woodland species. Common in suitable habitat throughout the region except in the south.
Call. A series of 'tu-tu-tu-tu' whistles which rise in pitch and end in a clear 'wheeoo-wheeoo'. *Afrikaans.* Witkoluil.

394 **Wood Owl** *Strix woodfordii* L 35 cm
This medium-sized owl can be identified by its lack of 'ear' tufts, its heavily barred brown underparts and pale, finely barred facial disc with large, dark brown eyes. Plumage coloration is variable, ranging from very dark brownish black to russet. *Habitat.* Thick evergreen and riverine forests, and exotic plantations. Common throughout except in the treeless central and western parts. *Call.* Close to the well-described 'tuwhit-towhoo' call but rendered as 'huoo-hoo-hoo', with the female's reply a higher pitched 'weooo'.
Afrikaans. Bosuil.

395 **Marsh Owl** *Asio capensis* L 36 cm
A nondescript, medium-sized owl with small 'ear' tufts situated above dark brown eyes. In flight shows pale buffish 'windows' on the primaries. When flushed during daytime will circle overhead before alighting. *Habitat.* Marshes and damp grasslands, avoiding thick reedbeds. Common throughout the region except in the dry central and western parts. *Call.* A harsh 'krikkk-krikkk'.
Afrikaans. Vlei-uil.

396 ▲   394 ▼   398 ▼   397 ▲   395 ▼   399 ▲

404 European Nightjar *Caprimulgus europaeus* L 25-28 cm
A large nightjar, it is paler than Freckled Nightjar and shows more white in the wing and tail. Lack of rufous on head and neck separates it from Fierynecked and Rufouscheeked nightjars. Reduced white on tail differentiates it from Natal and Mozambique nightjars, and it has much less white in the wing than Pennantwinged Nightjar. *Habitat.* Virtually any habitat except desert. Common summer visitor throughout except for the north-west. *Call.* Silent when in Africa. *Afrikaans.* Europese Naguil.

405 Fierynecked Nightjar *Caprimulgus pectoralis* L 24 cm
When seen at rest can be distinguished from Rufouscheeked Nightjar by its rich rufous, not orange-buff collar. Males not distinguishable in flight but females have off-white tips to outer tail feathers, a feature lacking in the female Rufouscheeked. *Habitat.* Found in a wide range of habitats and has adapted well to exotic plantations. Common except in the dry central and western areas. *Call.* A plaintive 'whue-whe-whe-whe-whe-whe', with the last notes faster and descending in pitch. *Afrikaans.* Afrikaanse Naguil.

406 Rufouscheeked Nightjar *Caprimulgus rufigena* L 24 cm
Differs from Fierynecked Nightjar by having an orange-buff, not rufous collar and by lacking the rufous on its breast. Female differs from the female Fierynecked by having no white on tail. *Habitat.* Dry thornveld and broad-leafed woodland, scrub desert and Karoo vegetation. The nightjar most common in the dry west. *Call.* A 'kow-kow-kow', a choking call note 'chukoo, chukoo', and a soft, purring sound. *Afrikaans.* Rooiwangnaguil.

407 Natal Nightjar *Caprimulgus natalensis* L 22 cm
A small nightjar, more buff in colour than any other nightjar in the region. Resembles the Mozambique Nightjar inasmuch as both have white outer tail feathers, but the Natal Nightjar is generally smaller and is paler in colour. *Habitat.* Palm savanna, usually in moist areas. Rare in the east, but more common in northern Botswana. *Call.* A 'chow-chow-chow' or 'chop-chop-chop' call. *Afrikaans.* Natalse Naguil.

408 Freckled Nightjar *Caprimulgus tristigma* L 28 cm
Similar in size to European and Pennantwinged nightjars but differs from both by its freckled greyish brown upperparts, which blend well with the rocky terrain it frequents. In flight distinguished from European Nightjar by having white inner, not outer tail feathers. *Habitat.* Rocky outcrops in woodland and hilly terrain. Recently found roosting on buildings in towns and cities. Common but localized throughout, except in the north-eastern, southern and central areas. *Call.* A yapping 'kow-kow-kow-kow'. *Afrikaans.* Donkernaguil.

409 Mozambique Nightjar *Caprimulgus fossii* L 24 cm
Resembles Natal Nightjar in having the outer tail feathers white, but differs by being darker brown, less buff above, and by its slightly larger size. Female distinguished from female Natal Nightjar by having less white on the tail, and by its larger size. *Habitat.* Coastal dune scrub and sandy woodland, often near lakes and rivers. Common in the north and east. *Call.* A gurgling, churring sound. *Afrikaans.* Laeveldnaguil.

410 Pennantwinged Nightjar *Macrodipteryx vexillaria* L 28 cm
Males are unmistakable: they show a broad white stripe across the primaries, and elongated inner primaries which trail well behind the bird. In flight, the broad white stripe gives the impression that the pennants and wing tips are disjointed from the rest of the wing. Female is a nondescript large nightjar with no white on wings or tail. *Habitat.* Open woodland and bushveld near hillsides or water. Summer visitor to the north and east. *Call.* A soft piping note and bat-like squeaking. *Afrikaans.* Wimpelvlerknaguil.

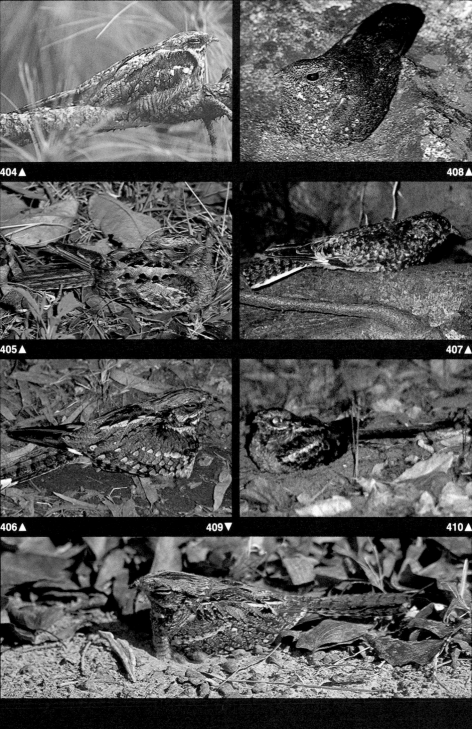

404 ▲

408 ▲

405 ▲

407 ▲

406 ▲

409 ▼

410 ▲

411 European Swift *Apus apus* L 17 cm
Difficult to distinguish from Pallid and Black swifts. Differentiated from the former only under optimum viewing conditions when both species are present: European Swift is darker, has far less white on its throat and has a more slender body. Differs from Black Swift by having secondaries and back uniform in colour. *Habitat.* Aerial, even sleeping on the wing at great heights. Common summer visitor throughout the region, except in the south. *Call.* Silent in Africa. *Afrikaans.* Europese Windswael.

412 Black Swift *Apus barbatus* L 18 cm
Can be distinguished (with difficulty) from European Swift if upperparts are seen clearly. The secondaries, especially the inner secondaries, are paler than the rest of the wing and back and show up as contrasting pale greyish brown areas. *Habitat.* Aerial. Traditional breeding sites are inland cliffs, but has been seen on Table Mountain. Common throughout except for central and north-western areas. *Call.* A high-pitched screaming at breeding sites. *Afrikaans.* Swartwindswael.

413 Bradfield's Swift *Apus bradfieldi* L 18 cm
Paler than both European and Black swifts. At close range and in good light a scaly, mottled effect is discernible on the underparts. In the field unlikely to be distinguished from Pallid Swift unless both are seen together. *Habitat.* Aerial and wide ranging, breeding on inland cliffs. Common only in the north-west. *Call.* High-pitched screaming at breeding sites. *Afrikaans.* Muiskleurwindswael.

414 Pallid Swift *Apus pallidus* L 17 cm
Very difficult to tell from Black, European and Bradfield's swifts unless seen together. Differs from Black and European by being paler with a more extensive white throat patch and paler forehead. Distinguished from Bradfield's Swift by being only slightly paler and lacking the scaled effect on the underparts. In comparison with all three species, it has a more robust body and a slower flight action. *Habitat.* Aerial. Rare vagrant with one record from the northern Cape. *Call.* Silent in Africa. *Afrikaans.* Bruinwindswael.

419 Mottled Swift *Apus aequatorialis* L 20 cm
Much the same size and shape as Alpine Swift but lacks the white underparts of that species. Has scaled and mottled underparts like Bradfield's Swift but is darker and much larger. *Habitat.* Aerial and wide ranging. Breeds on inland cliffs. Uncommon, found in the Eastern Highlands of Zimbabwe. Doubtful record from the south-western Cape. *Call.* The typical swift scream. *Afrikaans.* Bontwindswael.

418 Alpine Swift *Apus melba* L 22 cm
A large swift – its wingspan approaches that of the Rock Kestrel – showing white underparts and a dark breast band. Flight action is extremely powerful and fast. Often seen in mixed flocks with other swift species. *Habitat.* Aerial and wide ranging. Breeds on high inland cliffs. Common throughout but absent from the central area. *Call.* A loud trilling whistle. *Afrikaans.* Witpenswindswael.

411▲   412▼

414▲   419▼   413▼

418▼

420 Scarce Swift *Schoutedenapus myoptilus* L 17 cm
A nondescript, dull brown swift which has the body outline and tail shape of the Whiterumped Swift but lacks the white rump of that species. Distinguished from Black and European swifts by duller brown plumage and longer, more deeply forked tail. *Habitat.* Over cliffs and rocky bluffs in forested mountain regions. Rare in eastern Zimbabwe. *Call.* Not recorded.
Afrikaans. Skaarswindswael.

421 Palm Swift *Cypsiurus parvus* L 17 cm
The most slender and streamlined swift in the region. The long, thin wings, the elongated, deeply forked tail, and the grey-brown coloration are diagnostic. Occurs in small groups or in mixed flocks with other swifts. *Habitat.* Usually found in the vicinity of palm trees, it has also adapted to exotic palms in towns and cities. Common in the north and east. *Call.* A soft, high-pitched scream.
Afrikaans. Palmwindswael.

415 Whiterumped Swift *Apus caffer* L 15 cm
The long, deeply forked tail on this 'white-rumped' swift is diagnostic. The tail is frequently held closed, appearing long and pointed. The thin, white, U-shaped band across the rump is less obvious than in either the Little or Horus swifts. *Habitat.* Aerial, over open country, mountainous terrain and in towns and cities. Common summer visitor to all parts except central Botswana and the western regions. *Call.* Normally silent. *Afrikaans.* Witkruiswindswael.

416 Horus Swift *Apus horus* L 16 cm
Most likely to be confused with the Little Swift as both show a broad white rump, but this is a larger, more robust bird. It is also distinguished by its forked tail, which sometimes appears square-ended when closed. Differs from Whiterumped Swift by being dumpier, having more white on the rump and a less forked tail. *Habitat.* Aerial, frequently found over mountainous terrain, sandbanks and road cuttings. Uncommon and localized in the north, east and south. *Call.* Normally silent. *Afrikaans.* Horuswindswael.

417 Little Swift *Apus affinis* L 14 cm
The small size, large, square white rump and square tail are all diagnostic. In flight seems squat and dumpy, with wing ends that appear rounded. The common swift over cities and towns, often seen wheeling in tight flocks during display flights. *Habitat.* Aerial, found over towns, cities, bridges and cliffs. Common throughout the region except for central Botswana and the Namib desert. *Call.* Soft twittering and high-pitched screeching.
Afrikaans. Kleinwindswael.

422 Mottled Spinetail *Telacanthura ussheri* L 14 cm
Most likely to be confused with the Little Swift but differs in that the white throat extends on to the upper breast where it appears mottled, and by having a small white patch on the undertail coverts. The spines on the tail are evident only when the bird is in the hand. *Habitat.* Wide ranging but usually found in the vicinity of baobab trees. Uncommon and highly localized in the north-east. *Call.* Recorded as a soft twittering and a 'zi-zick'. *Afrikaans.* Gevlekte Stekelstert.

423 Böhm's Spinetail (Batlike Spinetail) *Neafrapus boehmi* L 9 cm
The smallest swift of the region. Superficially resembles white-bellied form of Brownthroated Martin but differs in flight action and by having a white rump. The white belly, square white rump and very short tail are diagnostic. Flight action is very fast and erratic, almost bat-like. *Habitat.* Thornveld, open broad-leafed woodland and in the vicinity of baobab trees. Uncommon and localized in the north-east. *Call.* Recorded as a high-pitched tri-tri-tri-peep.
Afrikaans. Witpensstekelstert.

172

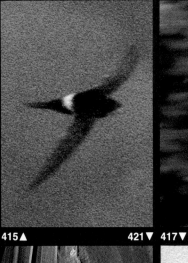

415▲   421▼   417▼   416▲

423▼   422▼

420▲

424 Speckled Mousebird *Colius striatus* L 35 cm
Distinguished from Whitebacked and Redfaced mousebirds by drabber brown coloration and black bill and face. Flight action is weaker and floppier than that of the other mousebird species. Often seen in small groups dashing from one bush to the next in 'follow-my-leader' fashion. *Habitat.* Thick tangled bush, fruiting trees in urban and suburban parks and gardens. Common in the north-east, east and south. *Call.* A harsh 'zhrrik-zhrrik'. *Afrikaans.* Gevlekte Muisvoël.

425 Whitebacked Mousebird *Colius colius* L 34 cm
In flight the back, with its white stripe bordered by black, is diagnostic. At rest differs from Speckled and Redfaced mousebirds by whitish bill, grey upperparts and red feet. *Habitat.* Thornveld, fynbos scrub, and semi-desert. Common endemic to the central, southern and dry western regions. *Call.* A whistling 'zwee-wewit'. *Afrikaans.* Witkruismuisvoël.

426 Redfaced Mousebird *Colius indicus* L 34 cm
Generally paler than Speckled and Whitebacked mousebirds and, when seen, the red face is diagnostic. In flight the grey rump contrasts slightly with the browner back and tail. The birds usually fly in small parties, and show a fast, powerful and direct flight action. *Habitat.* Thornveld, open broad-leafed woodland, and suburban gardens, avoiding forests and extremely dry regions. Common throughout in suitable habitat. *Call.* In flight a 'whee-whe-whe' is uttered, the first note being the highest in pitch. *Afrikaans.* Rooiwangmuisvoël.

427 Narina Trogon *Apaloderma narina* L 34 cm
A furtive species which, although brilliantly coloured, is very difficult to see as it normally sits with its back to the observer, well camouflaged by its leafy green surroundings. The combination of its crimson breast and bright emerald green body is diagnostic. The female lacks the green throat of the male and has a duller crimson breast. *Habitat.* Riverine and evergreen forests and dense, broad-leafed woodland. Common, but easily overlooked, in the north-east, east and south-east. *Call.* A soft, owl-like 'hooo-hooo'. *Afrikaans.* Bosloerie.

491 Angola Pitta *Pitta angolensis* L 23 cm
This brilliantly coloured bird is unmistakable but almost impossible to see in the dark understorey of the forests it frequents. Usually glimpsed as it rises off the forest floor in a bright flash of colour. In East Africa it feeds in association with the Giant Elephant Shrew, picking up insects exposed as the shrew overturns leaf litter. *Habitat.* Thick riverine forest and sandy coastal forest. A very rare and little-known species which breeds in the north-east. Vagrants are found further south. *Call.* Not recorded in our region. *Afrikaans.* Angolapitta.

490 African Broadbill *Smithornis capensis* L 14 cm
The male of this small, dumpy bird has a black cap and heavily streaked underparts. Female and imm. lack the black cap. The broad, flattened bill is not easily seen in the field except at close range and at certain angles. Only during the short, circular display flight is the white 'puffball' on the lower back fluffed out and visible. *Habitat.* Coastal forest and thickets, and the understorey of riverine forest. Uncommon and localized in the eastern, north-eastern and extreme north-central regions. *Call.* A frog-like 'prrrrrruup' is uttered during the display flight. *Afrikaans.* Breëbek.

424 ▲   426 ▼   491 ▼   425 ▲

427 ▼   490 ▼

429 Giant Kingfisher *Ceryle maxima* L 46 cm
The largest kingfisher in the region, the male is unmistakable with its long, heavy bill, its dark, white-spotted back and rufous breast. The female has a rufous belly. Does not hover over water as frequently as the Pied Kingfisher, preferring to sit on wires or branches overhanging water. *Habitat.* Wooded streams and dams, fast-flowing rivers in mountains, and coastal lagoons. Common but thinly distributed throughout, being absent from the dry west. *Call.* A loud, harsh 'kahk-kah-kahk'. *Afrikaans.* Reuse Visvanger.

428 Pied Kingfisher *Ceryle rudis* L 28 cm
The only kingfisher in the region with a pied black and white plumage. The male has a double breast band, the female a single, incomplete band. Frequently hovers over water before diving to seize a fish. *Habitat.* Any open stretch of fresh water, coastal lagoons and tidal rock pools. Common throughout but absent from some dry western areas. *Call.* A rattling twitter and a sharp, high-pitched 'chik-chik'. *Afrikaans.* Bontvisvanger.

430 Halfcollared Kingfisher *Alcedo semitorquata* L 20 cm
The black bill is diagnostic and is not seen on any other ad. small 'blue' kingfisher in the region. Larger than the Malachite Kingfisher, with which it overlaps in distribution, but lacks the turquoise crest and can be differentiated by the black bill character. Distinguished from the imm. Malachite Kingfisher (which does have a dark bill) by its larger size and the lack of the turquoise crest. *Habitat.* Wooded streams and coastal lagoons. Uncommon and thinly distributed in the north, east and south. *Call.* A high-pitched 'chreep' or softer 'peeek-peek'. *Afrikaans.* Blouvisvanger.

431 Malachite Kingfisher *Alcedo cristata* L 14 cm
Differs from similar, but smaller Pygmy Kingfisher by having the turquoise and black barred crown extending below the eye, and by lacking the violet wash on the sides of the head. Imm's black bill might lead to confusion with the Halfcollared Kingfisher, but this species is smaller, has a dark back and reddish brown underparts. *Habitat.* Reedbeds surrounding freshwater lakes, and along streams and lagoons. Common in suitable habitat throughout the region except the extreme dry west. *Call.* A high-pitched 'wheep-wheep' given in flight. *Afrikaans.* Kuifkopvisvanger.

432 Pygmy Kingfisher *Ispidina picta* L 13 cm
The smallest kingfisher in the region. Distinguished from similar Malachite Kingfisher by its smaller size, uniform blue crown which does not extend below the eye, and by a violet wash around the ear coverts. *Habitat.* Woodland thickets and coastal forests, often near water. Summer visitor in the south of its range. Common in the east and north-east. *Call.* A soft 'chip-chip' flight note. *Afrikaans.* Dwergvisvanger.

429▲   430▼   428▼

431▼   432▼

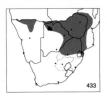

433 Woodland Kingfisher *Halcyon senegalensis* L 23 cm
Very similar to Mangrove Kingfisher but unlikely to be found in the same habitat. Differs by having a black lower mandible, not an all-red bill, and a much paler head with a black stripe extending from the base of the bill, through and behind the eye. *Habitat.* Thornveld, open broad-leafed woodland and riverine forest. Common summer visitor to the north-west, north-east and east. *Call.* A loud, piercing 'trrp-trrrrrrrrrr', the latter part descending. *Afrikaans.* Bosveldvisvanger.

434 Mangrove Kingfisher *Halcyon senegaloides* L 24 cm
Closely resembles the Woodland Kingfisher but is easily identified by its all-red, not red and black bill, and darker grey head. Differs from Greyhooded Kingfisher by having a blue back and lacking the chestnut belly, and can be distinguished from Brownhooded Kingfisher as that species has a black back, paler head and brown patches on the sides of the breast. *Habitat.* Mangrove swamps and well-wooded coastal rivers. Uncommon along the north-east and east coasts. *Call.* A noisy species in the mangroves, giving a loud ringing 'cheet choo-che che che', the latter part ending in a trill. *Afrikaans.* Mangliedvisvanger.

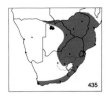

435 Brownhooded Kingfisher *Halcyon albiventris* L 24 cm
This species can be distinguished from other similar red-billed kingfishers by its brownish head streaked with black, rusty patches on the sides of the breast and well-streaked flanks. Differs from Striped Kingfisher by all-red, not black and red bill, and by the lack of a dark cap. *Habitat.* Thornveld, open broad-leafed woodland and coastal forests. Often found far from water. Has adapted to suburbia and is common in gardens and parks. Common in the north, east and south, being absent from the drier west. *Call.* A whistled 'kee-kee-kee-kee' and a harsher alarm note 'klee-klee-klee'. *Afrikaans.* Bruinkopvisvanger.

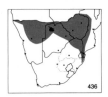

436 Greyhooded Kingfisher (Chestnutbellied Kingfisher)
Halcyon leucocephala L 20 cm
The grey head and chestnut belly are diagnostic. Distinguished from Brownhooded Kingfisher by the lack of any streaking on the head and flanks. The similar Mangrove Kingfisher has a heavier red bill and lacks the chestnut belly. *Habitat.* Mixed thornveld and miombo woodland. Uncommon summer visitor to the northern part of the region. *Call.* A loud, whistling 'cheeo cheeo weecho-trrrrr'. *Afrikaans.* Gryskopvisvanger.

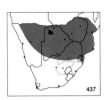

437 Striped Kingfisher *Halcyon chelicuti* L 18 cm
The dark cap lightly streaked with grey, the black and red bill and the white collar are diagnostic in this kingfisher. Distinguished from Brownhooded Kingfisher, with which it often associates, by its smaller size, darker capped appearance and the white collar. The blue on the back is evident only during flight. *Habitat.* Thornveld, riverine and coastal forests. Common in the northern, central and north-eastern regions. *Call.* A high-pitched, piercing 'cheer-cherrrrrr', the last notes running rapidly together. *Afrikaans.* Gestreepte Visvanger.

433 ▲   434 ▲

435 ▲   436 ▼ 437 ▼

438 European Bee-eater *Merops apiaster* L 28 cm
In our region, the only bee-eater with a chestnut crown and back. Imm. has a green back but its pale blue underparts should eliminate confusion with other bee-eaters in the region. *Habitat.* Thornveld, open broad-leafed woodland and adjacent grassy areas. Common summer visitor to the north with small populations in the west and south-west. *Call.* Far-carrying flight calls: 'prrrup' and 'krroop-krroop'. *Afrikaans.* Europese Byvreter.

440 Bluecheeked Bee-eater *Merops persicus* L 31 cm
Differs from Olive Bee-eater by having a green, not brown crown, a blue forehead, eyebrow stripe and cheeks, yellow throat and rusty upper breast. *Habitat.* Floodplains and adjacent broad-leafed woodland. Common but localized summer visitor to the north and north-east. *Call.* A liquid 'prrrup' and 'prrreo'. *Afrikaans.* Blouwangbyvreter.

441 Carmine Bee-eater *Merops nubicoides* L 36 cm
Ad. unmistakable. Imm. lacks the elongated central tail feathers, has a brown, not carmine back and has less brightly coloured underparts. *Habitat.* Colonial breeders in sandy banks, usually found near rivers or marshes and adjoining grasslands and bush. Common summer visitor to the north and north-east. *Call.* A deep 'terk, terk'. *Afrikaans.* Rooiborsbyvreter.

443 Whitefronted Bee-eater *Merops bullockoides* L 24 cm
The crimson and white throat, white forehead and lack of pointed tail projections are diagnostic. Imm. is a duller version of ad. Usually found in pairs. Form flocks to roost in trees or on rock ledges of riverbanks. *Habitat.* Wide, slow-moving rivers with steep sandbanks, and other freshwater expanses. Common in the north and north-east. *Call.* A 'qrrruk, qrrruk' and twittering noises when roosting. *Afrikaans.* Rooikeelbyvreter.

439 Olive Bee-eater *Merops superciliosus* L 31 cm
Differs from Bluecheeked Bee-eater by its brown, not green crown, rust-coloured throat and pale green underparts. Vaguely resembles Böhm's Bee-eater but has a dull brown, not chestnut cap, and is much larger. *Habitat.* Open broad-leafed woodland near lakes and swamps. Rare in the extreme north and north-east. *Call.* 'Prrrup'. *Afrikaans.* Olyfbyvreter.

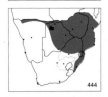

444 Little Bee-eater *Merops pusillus* L 17 cm
The smallest bee-eater in the region. Easily identified by its combination of small size, yellow throat, black collar, buff-yellow belly and square-ended, dark-tipped tail. In flight shows conspicuous russet underwings. *Habitat.* Forest edges in thornveld, and open clearings in coastal forest. Common in the east and north. *Call.* A 'zeet-zeet' or 'chip-chip'. *Afrikaans.* Kleinbyvreter.

445 Swallowtailed Bee-eater *Merops hirundineus* L 22 cm
The only fork-tailed bee-eater in the region. Most likely to be confused with Little Bee-eater but has a blue, not black collar, blue-green underparts and a blue, forked tail. Imm. shows the diagnostic forked tail, but lacks the yellow throat and blue collar. *Habitat.* Diverse: from semi-desert scrub to moist, evergreen forest. Common in the north. *Call.* A 'kwit-kwit' or soft twittering. *Afrikaans.* Swaelstertbyvreter.

442 Böhm's Bee-eater *Merops boehmi* L 21 cm
Differs from Olive Bee-eater in that it is much smaller, has a chestnut cap which extends down on to the nape, lacks the pale eyebrow stripe and has the russet throat spreading on to the breast. *Habitat.* Open areas in miombo woodland, wooded rivers and streams. Very rare, with only a few records from Moçambique. *Call.* A soft 'swee'. *Afrikaans.* Roeskopbyvreter.

138▲ 444▼ 445▼ 440▲ 442▼ 443▲

141▼ 439▼

446 European Roller *Coracias garrulus* L 31 cm
Differs from Lilacbreasted and Racket-tailed rollers by having an all-blue head and square-ended tail. Imm. Racket-tailed Roller appears similar when tail streamers are absent but that species has a distinctive white-streaked forehead, and a green, not blue crown and nape. *Habitat.* Thornveld, open grassland with scattered trees. Common summer visitor to all parts except the dry central and western districts. *Call.* Normally silent in our region but when alarmed will give a 'krack-krack' call. *Afrikaans.* Europese Troupant.

448 Racket-tailed Roller *Coracias spatulata* L 36 cm
Appears darker than Lilacbreasted Roller, has less blue in the wings, no lilac on the breast, and has elongated outer tail feathers with spatulate tips. Imm. lacks the diagnostic tail and differs from imm. Lilacbreasted Roller by having violet and brown wing coverts, and a blue, not lilac-tinged breast.
Habitat. Occurs in more moist broad-leafed woodland than Lilacbreasted Roller. Uncommon in the north-east. *Call.* Similar to that of Lilacbreasted Roller but higher pitched and with more cackling. *Afrikaans.* Knopsterttroupant.

447 Lilacbreasted Roller *Coracias caudata* L 36 cm
Distinguished from similar Racket-tailed Roller by its obvious lilac, not blue breast, by being generally paler and by having pointed, elongated outer tail feathers. Imm. lacks elongated outer tail feathers but the lilac-coloured breast differentiates it from the European and Racket-tailed rollers.
Habitat. Mixed areas of thornveld, open broad-leafed woodland and alongside roads. Common in the northern part of the region. *Call.* Harsh squawks and screams when displaying. *Afrikaans.* Gewone Troupant.

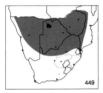

449 Purple Roller *Coracias naevia* L 38 cm
The largest roller of the region and easily identified by its broad, pale eyebrow stripe and lilac-brown underparts streaked with white. Imm. is a duller version of ad. *Habitat.* Dry thornveld and open broad-leafed woodland. Common but thinly distributed in the north. *Call.* In display flight utters a 'karaa-karaa'. *Afrikaans.* Groottroupant.

450 Broadbilled Roller *Eurystomus glaucurus* L 27 cm
The smallest roller in the region and the only one that appears dark with a bright yellow bill. In flight shows a blue tail and purple wing coverts. Imm., a duller version of ad., has greenish underparts streaked with black but also shows the bright yellow bill. *Habitat.* Riverine forest in thornveld. Shows particular preference for dead fig trees. *Call.* Harsh screams and cackles. *Afrikaans.* Geelbektroupant.

448 ▲

449 ▼

446 ▲

447 ▼

450 ▼

456 Silverycheeked Hornbill *Bycanistes brevis* L 75-80 cm
Noticeably far larger than Trumpeter Hornbill, with a huge whitish casque on top of the bill. The black on the underparts extends well down the belly, giving the impression of a much darker bird than the Trumpeter Hornbill. In flight distinguished from Trumpeter Hornbill by the lack of a white trailing edge to the wing. *Habitat.* Tall evergreen and riverine forests. Uncommon and localized in the extreme north-east. *Call.* A deeper wail than that of Trumpeter Hornbill, with an added harsh 'quark-quark'. *Afrikaans.* Kuifkopboskraai.

455 Trumpeter Hornbill *Bycanistes bucinator* L 58-65 cm
Similar to Silverycheeked Hornbill but is smaller, has a reduced and less obvious casque on top of its bill, and has a black throat and breast with the lower underparts white. If viewed at close range, it can be seen that the bare skin around the eye is pinkish red, not bluish green as in the Silverycheeked. *Habitat.* Evergreen, coastal and riverine forests. Common in the extreme north, east and south-east. *Call.* A wailing, infantile 'waaaaa-weeeee-waaaaa'. *Afrikaans.* Gewone Boskraai.

460 Crowned Hornbill *Tockus alboterminatus* L 54 cm
Differs from Bradfield's Hornbill by having a shorter red bill with a smaller casque and by having a darker head, back and tail. Distinguished from Monteiro's Hornbill by the lack of white in the wings and the small amount in the outer tail feathers. However, the range of the Crowned Hornbill does not overlap with either species except in the extreme north of Botswana, where it might be found with Bradfield's Hornbill. *Habitat.* Inland, coastal and riverine forests. Common in certain localities in the north, east and south-east. *Call.* A whistling 'chleeoo chleeoo'. *Afrikaans.* Gekroonde Neushoringvoël.

461 Bradfield's Hornbill *Tockus bradfieldi* L 56 cm
Differs from Monteiro's Hornbill by having no white in the wings and by lacking white outer tail feathers. Distinguished from Crowned Hornbill by being paler brown on the head and back, and by having a smaller red bill without a casque. *Habitat.* Open mopane woodland and mixed thornveld. Common in the extreme north-central region. *Call.* A whistling 'chleeoo' note, very similar to that of Crowned Hornbill. *Afrikaans.* Bradfieldse Neushoringvoël.

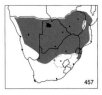

457 Grey Hornbill *Tockus nasutus* L 46 cm
In our region the male is the only small hornbill which has a dark bill. Female has the upper part of the bill yellow but differs from the Yellowbilled Hornbill by having a dark head and breast, and a broad white eyebrow stripe. Imm. resembles female. In flight this species shows a white stripe down the back. *Habitat.* Thornveld and dry broad-leafed woodland. Common in the northern part of the region. *Call.* A soft, plaintive whistling 'phee pheeoo phee pheeoo'. *Afrikaans.* Grysneushoringvoël.

455 ▲   460 ▼ 457 ▼   456 ▲   461 ▲

462 Monteiro's Hornbill *Tockus monteiri* L 56 cm
The large expanse of white on the outer tail feathers and the white patches on the secondaries are diagnostic in this large, red-billed hornbill. The similar Bradfield's Hornbill has no white in the wings and the white in the tail is confined to the tip. *Habitat.* Dry thornveld and broad-leafed woodland. Common endemic to north-western Namibia. *Call.* A hollow-sounding 'tooaak tooaak'. *Afrikaans.* Monteirose Neushoringvoël.

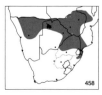

458 Redbilled Hornbill *Tockus erythrorhynchus* L 46 cm
The only small hornbill with an all-red bill. The pale head, broad white eyebrow stripe, and black and white speckled upperparts should obviate confusion with other larger, red-billed hornbills. Imm. has a shorter, less developed bill and buff, not white spotting on back and wing coverts. *Habitat.* Thornveld and mopane woodland. Common in the north and east but absent from the north-east. *Call.* A series of rapid 'wha wha wha wha wha' calls followed by a 'kukwe kukwe' note. *Afrikaans.* Rooibekneushoringvoël.

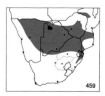

459 Yellowbilled Hornbill *Tockus flavirostris* L 55 cm
Very similar to Redbilled Hornbill in plumage coloration but has a diagnostic large yellow bill. Female Grey Hornbill has a shorter, part yellow bill but has a dark head and throat. *Habitat.* Thornveld and dry broad-leafed woodland. Common in the north and east but absent from a large part of Zimbabwe and Moçambique. *Call.* A rapid, hollow-sounding 'tok tok tok tok tok toka toka toka'. *Afrikaans.* Geelbekneushoringvoël.

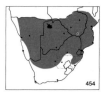

454 Scimitarbilled Woodhoopoe *Phoeniculus cyanomelas* L 26 cm
Smaller than the Redbilled and Violet woodhoopoes, this species differs from their imms. (which also have black bills) by its long, extremely decurved bill and black, not red legs and feet. In the field it appears black, except in direct sunlight when a purple sheen is noticeable. *Habitat.* Dry thornveld and open broad-leafed woodland. Common in the northern part of the region.
Call. High-pitched whistling 'sweep-sweep-sweep' and a harsher chattering. *Afrikaans.* Swartbekkakelaar.

452 Redbilled Woodhoopoe *Phoeniculus purpureus* L 36 cm
Larger than the Scimitarbilled Woodhoopoe, this species has a long, decurved red bill, red legs and a long, white-tipped tail. Imm. has a black bill but it is far less decurved than that of the Scimitarbilled. In good light, the bottle-green head and back of this species distinguish it from the Violet Woodhoopoe. *Habitat.* A wide variety of woodland and thornveld habitats. Common in the north, east and south-east. *Call.* Harsh chattering and cackling calls, usually uttered by groups of birds. *Afrikaans.* Gewone Kakelaar (Rooibekkakelaar).

453 Violet Woodhoopoe *Phoeniculus damarensis* L 40 cm
Easily confused with Redbilled Woodhoopoe but, in good light, this species' violet, not bottle-green head and back can be seen. Noticeably larger, with a floppier flight action. Imm. is indistinguishable from imm. Redbilled in the field. *Habitat.* Dry thornveld, wooded dry watercourses and mopane woodland. Common endemic to northern Namibia. *Call.* Harsh cackling indistinguishable from that of Redbilled Woodhoopoe. *Afrikaans.* Perskakelaar.

451 Hoopoe *Upupa epops* L 28 cm
Unmistakable with its long, decurved bill and long, black-tipped crest which is held erect when the bird is alarmed. Its cinnamon-coloured body, and black and white barred wings and tail are conspicuous in flight. *Habitat.* Thornveld, open broad-leafed woodland, parks and gardens. Common throughout the region except in desert areas. *Call.* A frequently uttered 'hoop-hoop-hoop'. *Afrikaans.* Hoephoep.

459 ▲   454 ▼   453 ▲   452 ▲

458 ▼   462 ▼   451 ▼

474 Greater Honeyguide *Indicator indicator* L 20 cm
The male's pink bill, dark crown, black throat and white ear patches are diagnostic. The female is fairly nondescript but differs from Scalythroated Honeyguide by having a clear, unmarked breast and by showing a white rump patch in flight. Imm. has a yellow wash across the breast. *Habitat.* Thornveld and open broad-leafed woodland, avoiding forests. Common but thinly distributed in the north, east and south-east. Absent from the dry western sector. *Call.* A 'whit-purr, whit-purr' note or, when agitated, a harsh rattling chatter. *Afrikaans.* Grootheuningwyser.

475 Scalythroated Honeyguide *Indicator variegatus* L 19 cm
The only honeyguide in the region to have the throat and breast mottled and speckled. Lacks the pale rump of the slightly larger Greater Honeyguide. When seen in the forest canopy, most easily confused with Lesser Honeyguide but the lack of moustachial stripes and greenish wash on wings, together with its more obvious scaly, mottled throat, should identify it. *Habitat.* Forests, thick broad-leafed woodland, and tall trees in riverine forests. Uncommon and thinly distributed in the north-east, east and south-east. *Call.* A difficult-to-locate, ventriloquistic 'trrrrrrrr' is given from a concealed call-site, usually high in a tall tree. *Afrikaans.* Gevlekte Heuningwyser.

476 Lesser Honeyguide *Indicator minor* L 15 cm
Intermediate in size between the Greater and Sharpbilled honeyguides. Differs from the latter by its much thicker bill, dark moustachial stripes and greenish wash on the wings. Distinguished from Greater Honeyguide by lacking the black throat, by having a dusky breast and lacking white on the rump. *Habitat.* Wide range of woodland, forests and thornveld and has adapted to urban and suburban gardens. Common, but easily overlooked, in the southern, eastern, northern, central and north-western regions. *Call.* Emits a far-carrying 'klew klew' from a regular call-site in a leafy canopy. *Afrikaans.* Kleinheuningwyser.

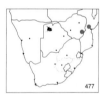

477 Eastern Honeyguide *Indicator meliphilus* L 13 cm
The short, thick bill should help eliminate confusion with Slenderbilled and Sharpbilled honeyguides. Much smaller than the Lesser Honeyguide and greener on the upperparts, especially the head. It is also greener on the crown, nape and back than the similar-sized Slenderbilled Honeyguide. *Habitat.* A forest species which frequents tall trees. Fewer than five records from Moçambique and eastern Zimbabwe. *Call.* Not recorded in our region but is reported to utter a thin whistle. *Afrikaans.* Oostelike Heuningwyser.

478 Sharpbilled Honeyguide *Prodotiscus regulus* L 13 cm
Similar to Eastern and Slenderbilled honeyguides but is more slender, with a longer tailed appearance and looks more like a flycatcher than a honeyguide. Differs from both species by the lack of any greenish wash on the upperparts, and by having a white throat and grey wash across the breast. *Habitat.* Thornveld, forest edges and exotic plantations. Uncommon and thinly distributed in the northern, north-eastern and eastern parts of the region. *Call.* A soft, thin 'tseep' or a tinkling 'trrrr'. *Afrikaans.* Skerpbekheuningvoël.

479 Slenderbilled Honeyguide *Prodotiscus zambesiae* L 12 cm
Distinguished from Sharpbilled Honeyguide by its stockier build, shorter tailed appearance and by having a greenish wash across its back and rump. Similar in shape to Eastern Honeyguide but differs by having a very small, thin bill and a grey, not greenish crown and nape. *Habitat.* Restricted to miombo woodland. Uncommon in the extreme north of Botswana and Zimbabwe, with vagrant records in Moçambique. *Call.* A repeated 'skeea, skeea' is uttered in display flight. *Afrikaans.* Dunbekheuningvoël.

479 ▲   476 ▼   477 ▲   478 ▲

475 ▼ 474 ▼

473 Crested Barbet *Trachyphonus vaillantii* L 23 cm
The orange face, erectile shaggy crest, and yellow underparts with broad black breast band are diagnostic and render this, the largest barbet of the region, totally unmistakable. *Habitat.* Open broad-leafed woodland, riverine forest and thornveld. Has adapted to urban and suburban gardens. Common in the central, northern and eastern parts of the region. *Call.* A long, sustained trilling 'trrrrrrrr . . .', much like a muted alarm clock.
Afrikaans. Kuifkophoutkapper.

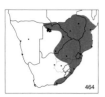

464 Blackcollared Barbet *Lybius torquatus* L 20 cm
The bright red face and throat, broadly bordered with black, are diagnostic. Uncommon variant has the red on face and throat replaced with yellow. Imm. has head and throat dark brown, streaked with orange and red. *Habitat.* Thornveld, dry broad-leafed woodland and coastal forest. Common in suitable habitat and has adapted to parks and gardens. Found in the north and east. *Call.* Far-carrying duet starting off with a harsh 'krrr krrrr' and exploding into a 'tooo puudly tooo puudly', the 'tooo' being higher pitched. *Afrikaans.* Rooikophoutkapper.

466 White-eared Barbet *Stactolaema leucotis* L 17 cm
The white ear-stripes and belly contrast with the dark brown to black head, throat and back. Imm. very similar to ad. but has paler base to bill and slightly paler back and throat. *Habitat.* Coastal forest and bush, especially alongside rivers. Common in the east and north-east. *Call.* A loud twittering 'treee treeetee teeetree' and various harsher 'waa waa' notes.
Afrikaans. Witoorhoutkapper.

467 Whyte's Barbet (Yellowfronted Barbet) *Stactolaema whytii* L 18 cm
Generally paler than White-eared Barbet, lacking the white ear patches of that species. The pale yellow forehead, white stripe below the eye and whitish patches on the wings eliminate confusion with any other barbet in the region. Imm. has no yellow on forehead and has a paler head and throat than the ad. *Habitat.* Miombo woodland and riverine forest, showing a preference for fig trees. Common, but very localized, in the north-east. *Call.* A soft 'coo' is repeated several times. *Afrikaans.* Geelbleshoutkapper.

468 Woodward's Barbet (Green Barbet) *Cryptolybia woodwardi* L 17 cm
A drab, green barbet with a dark crown and pale green ear patches. In flight shows pale green areas at the base of the primaries. Green Tinker Barbet is vaguely similar but confusion is unlikely as their ranges never overlap and Woodward's is much larger and has no yellow rump. *Habitat.* Canopy of coastal forest. Restricted to, but common in the Ngoye forest, Zululand. *Call.* A hollow-sounding 'kwop-kwop-kwop' is repeated many times. *Afrikaans.* Groenhoutkapper.

464 ▲

468 ▼

467 ▲

466 ▼

473 ▼

465 Pied Barbet *Lybius leucomelas* L 18 cm
The only large black and white barbet in the region. Has a red forehead, yellow in front of the eye and a broad white stripe behind the eye. Differs from the smaller Redfronted Tinker Barbet by having white underparts and a black bib. *Habitat.* Dry broad-leafed woodland, and thornveld. Common throughout the region except in the north-east and south-east. *Call.* A nasal 'nehh-nehh' and a deep 'doh-doh-doh'. *Afrikaans.* Bonthoutkapper.

469 Redfronted Tinker Barbet *Pogoniulus pusillus* L 11 cm
Differs from the Yellowfronted Tinker Barbet only in the colour of the forehead which, in this species, is bright red. The imm. of these two species are not reliably distinguishable in the field. *Habitat.* In comparison with the Yellowfronted Tinker Barbet, prefers more moist broad-leafed woodland and forests. Common in coastal regions of the east and south-east.
Call. A continual, monotonous 'konk konk konk', repeated in rapid succession. *Afrikaans.* Rooiblestinker.

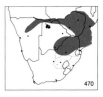

470 Yellowfronted Tinker Barbet *Pogoniulus chrysoconus* L 11 cm
The yellow forehead is the only feature that distinguishes this species from the Redfronted Tinker Barbet. Considerable variation occurs in the forehead coloration: it ranges from pale yellow to bright orange but never attains the bright red of the Redfronted Tinker Barbet. *Habitat.* Dry broad-leafed woodland and thornveld. Common in the central and north to north-eastern regions. *Call.* Very similar to that of Redfronted Tinker Barbet but is recorded as being lower in pitch. *Afrikaans.* Geelblestinker.

471 Goldenrumped Tinker Barbet *Pogoniulus bilineatus* L 10 cm
This small barbet has a diagnostic black crown with white stripes on the sides of the head, and a black back. The small yellow rump patch is not easy to see in the field as the bird is a canopy feeder. Imm. has the black upperparts narrowly barred and spotted with yellow. *Habitat.* Evergreen, coastal and riverine forests. Common in the lowlands of the north-east and east.
Call. 'Doo-doo-doo-doo', a lower pitched, more ringing note than that of Redfronted Tinker Barbet, repeated in phrases of four to six, not continuously. *Afrikaans.* Swartblestinker.

472 Green Tinker Barbet *Pogoniulus simplex* L 10 cm
This small, uniformly drab green barbet is unmistakable with its bright yellow rump and pale yellow panel showing on the folded wing. Lacks any diagnostic head markings. *Habitat.* Found in the canopy of tall, coastal forests. Rare, with only one positive record south of Beira, Moçambique. *Call.* Not recorded within our region but reported to be a 'pop-op-op-op-op-op' ringing note, not unlike that of other small tinker barbets. *Afrikaans.* Groentinker.

472▲ 470▼ 469▼

465▼

471▼

489 Redthroated Wryneck *Jynx ruficollis* L 19 cm
The brown barred and mottled plumage should eliminate confusion with any woodpecker. Might be mistaken for Spotted Creeper but is much larger and has a dark chestnut throat and breast. Shows jerky movements, similar to those of woodpeckers, as it creeps around branches and tree trunks. *Habitat.* Thornveld and dry broad-leafed woodland. Has adapted well to suburban gardens and exotic stands of eucalyptus. Common, but localized in the south-eastern and central regions. *Call.* A loud, piercing 'kwik-kwik-kwik'. *Afrikaans.* Draaihals.

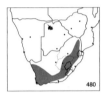

480 Ground Woodpecker *Geocolaptes olivaceus* L 25 cm
The only entirely terrestrial woodpecker in the region. Easily identified by its diagnostic pinkish red belly and rump. Female and imm. are very similar to the male but have much reduced and duller pinkish red on belly and rump, and lack the reddish moustachial stripes. *Habitat.* Boulder-strewn grassy hill slopes, mountain gullies and deep, dry dongas. Common endemic to the south and east. *Call.* A far-carrying 'kee-urrr, kee-urrr' and a sharper 'kleek'. *Afrikaans.* Grondspeg.

488 Olive Woodpecker *Mesopicos griseocephalus* L 20 cm
The greyish head, unmarked dull green body and red rump are diagnostic. Female and imm. have uniformly grey heads whereas the male has a bright red hind crown and nape. *Habitat.* Evergreen and riverine forests and thick coastal scrub. Common in the southern, south-eastern, eastern and the extreme north-central regions. *Call.* A bulbul-like 'chweet, chweet, chereep, chereep'. *Afrikaans.* Gryskopspeg.

481 Bennett's Woodpecker *Campethera bennettii* L 24 cm
Male differs from all other woodpeckers in the region by having an all-red forehead, crown and moustachial stripes. Female readily identifiable by her brown throat and stripe below the eye. *Habitat.* Dry thornveld and open broad-leafed woodland. Uncommon and thinly distributed in the north. *Call.* A high-pitched, chattering 'whirrr-trrr-whrrr-itt', often uttered in a duet. *Afrikaans.* Bennettse Speg.

482 Specklethroated Woodpecker *Campethera b. scriptoricauda* L 20 cm
A description is not included because this bird is now considered a subspecies of Bennett's Woodpecker. (Short, 1982: *Woodpeckers of the World.* Delaware Museum of Natural History: Greenville.) *Afrikaans.* Tanzaniese Speg.

480 ▲   481 ▼   488 ▼   489 ▲

483 Goldentailed Woodpecker *Campethera abingoni* L 23 cm
Very similar to the Knysna Woodpecker but is paler below and much less heavily streaked and blotched with brown on breast and belly. The male has a black and red forehead, thus eliminating confusion with Bennett's Woodpecker. *Habitat.* Thornveld, dry, open broad-leafed woodland and coastal forest. Common in the eastern and northern sectors of the region. *Call.* A loud, nasal 'wheeeeeaa' shriek. *Afrikaans.* Goudstertspeg.

484 Knysna Woodpecker *Campethera notata* L 20 cm
Both male and female are very much darker than Goldentailed Woodpecker and are heavily spotted, not streaked, with dark brown on the underparts. The male's dark red forehead, crown and moustachial stripes are heavily blotched with black, making them less conspicuous than those of the Goldentailed Woodpecker. Female has indistinct moustachial stripes. *Habitat.* Coastal evergreen forests. Common endemic to the south and south-east in suitable habitat. *Call.* A higher pitched shriek than that of Goldentailed Woodpecker. *Afrikaans.* Knysnaspeg.

485 Little Spotted Woodpecker *Campethera cailliautii* L 16 cm
Male distinguished from similar-sized Cardinal Woodpecker by its red forehead and crown, by a yellow-spotted, not barred back and by the lack of any moustachial stripes. The female differs from the female Cardinal Woodpecker by having a red hind crown and by lacking black moustachial stripes. *Habitat.* Evergreen forests and open broad-leafed woodland. Uncommon in the north-east. *Call.* Described as a shrill 'hee-hee-hee-hee'. *Afrikaans.* Gevlekte Speg.

486 Cardinal Woodpecker *Dendropicos fuscescens* L 15 cm
Could be confused with Little Spotted Woodpecker only in size. Otherwise differs markedly in that it has bold black moustachial stripes and appears black and white, not green and white, all over. Male can also be distinguished from male Little Spotted Woodpecker by its brown, not red forehead. *Habitat.* Frequents a wide range of habitats, from thick forest to dry thornveld. Common throughout except in extreme desert and mountainous regions. *Call.* Taps wood very rapidly and quietly. A high-pitched 'krrrek krrrek krrrek' shriek is uttered. *Afrikaans.* Kardinaalspeg.

487 Bearded Woodpecker *Thripias namaquus* L 25 cm
This large woodpecker has a white face with bold black moustachial stripes and a black stripe through and behind the eye. Male has a red hind crown and nape (female's is black), and both sexes have very dark underparts which are finely barred black and white. *Habitat.* Occurs in dry broad-leafed woodland and riverine forests, especially in areas where there are many dead trees. Common, but localized in the north. *Call.* Extremely loud wood tapping. A loud 'kweek-eek-eek-eek' is given. *Afrikaans.* Baardspeg (Namakwaspeg).

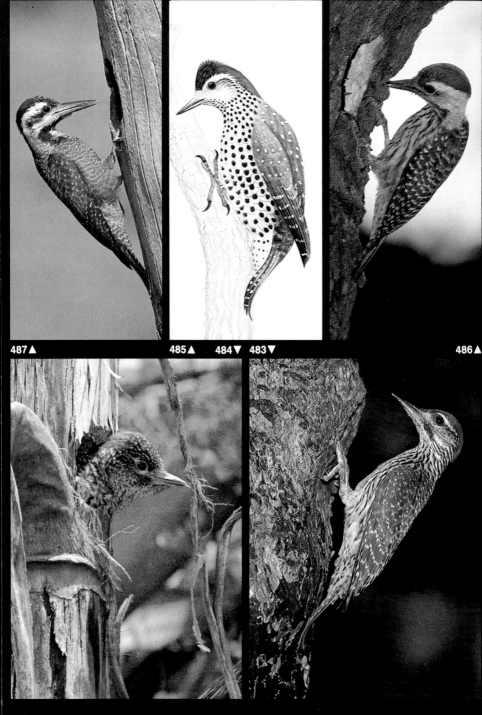

487▲   485▲   484▼   483▼   486▲

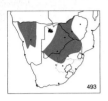

493 Monotonous Lark *Mirafra passerina* L 14 cm
The white flanks and belly distinguish this species from the very similar Melodious Lark. Where the range of this species overlaps with that of Stark's Lark, this species can be identified in flight by its chestnut wing patches. Display flight is low-level and short, with the bird launching itself from a perch, all puffed up as it sings. *Habitat.* Thornveld and mopane woodland with sparse grass cover. Common but nomadic endemic to the central and north-western regions. *Call.* Frequently repeated 'trrp-chup-chip-choop', during day and night. *Afrikaans.* Bosveldlewerik.

495 Clapper Lark *Mirafra apiata* L 15 cm
Very difficult to distinguish from Flappet Lark unless seen in display flight or heard calling. Generally much more rufous than Flappet Lark, with white, not buff outer tail feathers. Display flight comprises a steep climb during which the bird rattles its wings. This is followed by a rapid descent, with legs trailing, during which the bird calls. This sequence may begin from the ground, or during high-level flight. *Habitat.* Upland grasslands, fynbos and open Kalahari scrub. Common endemic to the southern, central and north-western regions. *Call.* A long drawn-out whistle, 'pooooeeee', preceded by loud wing rattling. *Afrikaans.* Hoëveldklappertjie.

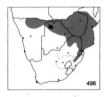

496 Flappet Lark *Mirafra rufocinnamomea* L 15 cm
Plumage coloration and markings are very similar to Clapper Lark but this species is usually very much greyer, less rufous and has buff, not white outer tail feathers. The display flight is usually very high, with loud wing rattling. *Habitat.* Lowland grasslands, open thornveld and broad-leafed woodland. Common in the north-east and in northern Botswana. *Call.* A short 'tuee' call given when perched. During display flight wings are rattled in a series of short bursts. *Afrikaans.* Laeveldklappertjie.

505 Dusky Lark *Pinarocorys nigricans* L 19 cm
This large lark is more likely to be confused with a Groundscraper Thrush (p. 224) than any other lark species. The bold black and white face pattern, heavy spotting on underparts and its habit of perching in trees, all add to its thrush-like appearance. It has the unusual habit of raising its wings slightly above body level when walking. *Habitat.* Open grassy areas in thornveld and broad-leafed woodland. Frequently found on newly burnt grassland. Uncommon summer visitor from tropical Africa to the northern part of the region. *Call.* When flushed it utters a soft 'chrrp, chrrp'. *Afrikaans.* Donkerlewerik.

513 Bimaculated Lark *Melanocorypha bimaculata* L 17 cm   □
A large, dumpy lark with a heavy, pale bill and diagnostic black patches on sides of breast. In flight the short dark tail, narrowly tipped with white, can be seen. *Habitat.* Stony desert and grasslands. One record of doubtful origin from coastal Namibia. *Call.* Not recorded in our region. *Afrikaans.* Swartboslewerik.

512 Thickbilled Lark *Galerida magnirostris* L 18 cm
The thick-based, heavy bill, with yellow at the base of the lower mandible, is diagnostic. A robust, heavily built lark with a relatively short tail and boldly streaked underparts. Has a noticeable crest which is raised when the bird is alarmed or singing. *Habitat.* Lowland and montane fynbos, Karoo scrub regions and wheatfields. Common endemic to the south-western regions. *Call.* A soft but far-carrying 'treeeleeeleee', likened to a rusty gate being opened. *Afrikaans.* Dikbeklewerik.

96▲ 505▼

495▲ 512▼

493▼

494 Rufousnaped Lark *Mirafra africana* L 18 cm
The bright red patches on the wings and the rufous nape (when seen) are
diagnostic. Differs from Shortclawed by its heavier bill, and rufous nape and
wing patches. *Habitat.* Diverse: from open grasslands with stunted bushes to
thornveld and cultivated areas. Common throughout except in the south and
dry west. *Call.* A frequently repeated 'treelee-treelooe' when perched. In
display flight, a jumbled mixture of imitated calls. *Afrikaans.* Rooineklewerik.

500 Longbilled Lark *Certhilauda curvirostris* L 20 cm
A large lark with a diagnostic long, decurved bill and a long thin tail. Very
variable in colour. Distinctive display flight in which the bird rises sharply from
the ground and descends gradually, giving its duosyllabic call.
Habitat. Grasslands, fynbos, Karoo scrub and stony ridges. Common endemic
to the southern, western and central regions. *Call.* A prolonged 'cheeeee-
ooooop' during the display flight. *Afrikaans.* Langbeklewerik.

501 Shortclawed Lark *Certhilauda chuana* L 19 cm
Differs from Rufousnaped by lacking any rufous on the nape or wings but, in
flight, it does show a rust-coloured rump. The buff-white eyebrow stripe runs
directly from the base of the straight, slender bill to the nape, giving a slightly
capped effect. *Habitat.* Dry thornveld along rivercourses and open grassy
areas. Uncommon endemic to the northern Cape, southern Botswana and
western Transvaal. *Call.* A short 'chreep-chuu-chree', given when perched in a
tree. Display flight similar to Longbilled Lark's. *Afrikaans.* Kortkloulewerik.

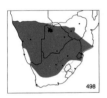

498 Sabota Lark *Mirafra sabota* L 15 cm
This small, greyish lark lacks the chestnut seen in the wings of the
Monotonous, Fawncoloured and Melodious larks. The bill is short and dark
(the lower mandible being paler), and a straight white eye-stripe runs from the
base of the bill to the nape, giving the head a capped effect. *Habitat.* Dry
thornveld and open broad-leafed woodland, often in rocky areas. Common
except in the south and north-east. *Call.* A jumbled song of rich, melodious
'chips' and twitterings. Mimics other birds. *Afrikaans.* Sabotalewerik.

502 Karoo Lark *Certhilauda albescens* L 17 cm
In the south, it is greyish brown with a noticeable white stripe above and below
the eye, and has dark ear coverts. In flight, shows a very dark-coloured tail. In
the north, differs from Red Lark by its shorter, more slender bill and rufous
upperparts streaked with dark brown. Distinguished from Dune Lark by having
darker rufous upperparts and more boldly streaked underparts.
Habitat. Fynbos, dry Karoo scrub and wheatlands. Common endemic to the
south-west. *Call.* A short 'chleeep-chleeep-chrrr-chrrrp' song is given in low-
level display flight. *Afrikaans.* Karoolewerik.

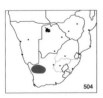

504 Red Lark *Certhilauda burra* L 19 cm
Differs from Dune Lark by rich rufous upperparts, and boldly streaked and
spotted underparts. Similar to red form of Karoo Lark but is larger with a much
heavier bill and has plain rufous upperparts. Has a very upright stance.
Habitat. Scrub-covered red sand dunes. Rare and highly localized endemic to
the northern Cape. *Call.* When flushed gives a short 'chrrk'. Song described as
a short 'toodly-woo tu-wee'. *Afrikaans.* Rooilewerik.

503 Dune Lark *Certhilauda erythrochlamys* L 17 cm
Distinguished from Karoo Lark by paler colour, plain or slightly streaked rufous
back, and slightly streaked underparts. Differs from Red Lark by longer, more
slender bill, less bold spotting on underparts and much paler rufous
upperparts. *Habitat.* Scrub growth on gravel plains in the Namib desert.
Uncommon endemic. *Call.* Similar to Karoo Lark's. *Afrikaans.* Duinlewerik.

200

494 ▲

500 ▼

502 ▲

504 ▼

501 ▼

503 ▼

498 ▼

506 Spikeheeled Lark *Chersomanes albofasciata* L 15 cm
The long straight bill, white throat patch contrasting strongly with the lower underparts, and the underside of the short tail narrowly tipped with white, are diagnostic. Has a very upright stance. *Habitat.* Sparse grasslands, scrub desert and desert gravel plains. Common in the central and dry western regions. *Call.* A trilling 'trrrep, trrrep' flight call. *Afrikaans.* Vlaktelewerik.

509 Botha's Lark *Calandrella fringillaris* L 12 cm
Difficult to distinguish from Pinkbilled Lark, however, this species has a slender, dark-tipped pink bill, which is thinner at the base. It is paler below with less contrast between the white throat and breast, and has slightly streaked buffy flanks. *Habitat.* Heavily grazed upland grasslands and grassy areas on stony ridges. Uncommon endemic, highly localized in the south-eastern Transvaal. *Call.* In flight utters a 'chuk, chuk'. *Afrikaans.* Vaalrivierlewerik.

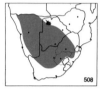

508 Pinkbilled Lark *Calandrella conirostris* L 13 cm
The short, conical, pink bill is diagnostic. Differs from Botha's Lark by its more robust, all-pink (not dark-tipped) bill, darker underparts which contrast boldly with the white throat, and clear, not streaked flanks. Where their ranges overlap, it can be differentiated from Stark's Lark by being less grey, lacking any obvious crest and having no dark tip to the bill. *Habitat.* Upland grasslands, farmlands and desert scrub. A common but nomadic endemic to the central and north-western regions. *Call.* When flushed, small flocks utter a soft 'si-si-si-si'. *Afrikaans.* Pienkbeklewerik.

499 Rudd's Lark *Heteromirafra ruddi* L 14 cm
Small, it appears large headed with a short, very thin tail. If clearly seen, the buff stripe down the centre of the crown is diagnostic. Display flight is high, with the bird often hanging into the wind whilst singing. *Habitat.* Upland grasslands, usually near damp depressions. Uncommon and highly localized endemic to the Transvaal and Orange Free State. *Call.* A clear, whistled song, 'pee-witt-weerr', is given in flight. *Afrikaans.* Drakensberglewerik.

507 Redcapped Lark *Calandrella cinerea* L 16 cm
Coloration very variable but it retains the diagnostic rufous cap and smudges on sides of breast. The underparts are white to off-white and unmarked, unlike any other lark in the region except Gray's, which lacks the red cap and shoulder smudges. *Habitat.* Ranges from true desert to eastern grasslands. Common but localized, being absent from areas in the north-east.
Call. A sparrow-like 'tchweerp' given in flight. Song is a sustained jumble of melodious phrases given during display flight. *Afrikaans.* Rooikoplewerik.

492 Melodious Lark (Singing Bush Lark) *Mirafra cheniana* L 12 cm
The white throat contrasts markedly with the lower underparts as the flanks and belly are buffish, not white as in Monotonous Lark. Display flight is characteristic: the bird rises to a great height and circles on whirring wings, singing all the while. *Habitat.* Gently sloping areas in upland grasslands. Common but localized endemic to the central eastern districts. *Call.* Call-note is a 'chuk chuk chuer', with a jumbled melodious song. Mimics other birds. *Afrikaans.* Spotlewerik.

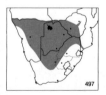

497 Fawncoloured Lark *Mirafra africanoides* L 14 cm
Variable, but usually a plain, buff-colour with well-marked upperparts, white underparts, and a slightly streaked breast. Differs from Sabota by less boldly streaked underparts and by noticeable white stripe below the eye.
Habitat. Kalahari scrub, and thornveld and dry bushveld in the west. Common in the central and northern parts. *Call.* A jumble of harsh 'chips' and twitterings, ending in a buzzy slur, given when perched on a bush top or during the short, fluttering display flight. *Afrikaans.* Vaalbruinlewerik.

499▲

506▲

509▲

507▲

492▲

508▼

497▲

511 **Stark's Lark** *Calandrella starki* L 14 cm
Much paler than Sclater's and Pinkbilled larks, this species lacks the teardrop mark of the former and differs from both by having a long erectile crest. Bill is a pale flesh colour and has a dark tip, distinguishing it from Pinkbilled Lark. Sometimes occurs in very large flocks. *Habitat.* Stony desert scrub to gravel plains of the Namib desert. A common but localized endemic to Namibia and the northern Cape. *Call.* Flight call is a short 'chree-chree'. Song given during display flight is a melodious jumble of notes. *Afrikaans.* Woestynlewerik.

510 **Sclater's Lark** *Calandrella sclateri* L 14 cm
When seen at close range, a dark brown teardrop mark below the eye is both obvious and diagnostic. In flight differs from Stark's and Pinkbilled larks by its bold white outer tail feathers which broaden towards the tail base and which, in profile, suggest a pale rump area. *Habitat.* Stony semi-desert with stunted Karoo scrub. Uncommon and localized endemic to the central dry west. *Call.* Flight call is a repeated 'tchweet-tchweet'. *Afrikaans.* Namakwalewerik.

514 **Gray's Lark** *Ammomanes grayi* L 14 cm
This small lark is the palest of the region and is unlikely to be confused with any other lark species. The pale desert form of the Tractrac Chat is similar in colour but that species has an obvious white base to its tail and a very upright stance on long legs. *Habitat.* Stony gravel plains and watercourses, and along the coastal desert strip of Namibia. Uncommon and difficult-to-locate endemic. *Call.* Flight call is a short 'tseet' or 'tew-tew'. *Afrikaans.* Namiblewerik.

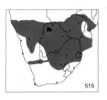

515 **Chestnutbacked Finchlark** *Eremopterix leucotis* L 12 cm
Differs from Greybacked Finchlark by having a chestnut back and forewings and a black crown. Female differs from female Greybacked Finchlark by having chestnut wing coverts. In flight shows white outer tail feathers, a feature which the other finchlarks lack. *Habitat.* Road verges and cultivated lands, sparsely grassed parts of thornveld, and lightly wooded areas. Common but nomadic in the central and northern regions. *Call.* Flocks in flight utter a short 'chip-chwep'. *Afrikaans.* Rooiruglewerik.

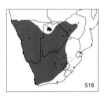

516 **Greybacked Finchlark** *Eremopterix verticalis* L 13 cm
The greyish back and wings distinguish this species from the Chestnutbacked Finchlark. The male is further distinguished from the male Chestnutbacked by a white patch on the hind crown. Female is very much greyer in appearance than the female Chestnutbacked Finchlark. *Habitat.* Cultivated lands, and scrub and true desert. A common but nomadic endemic to the central and dry western regions. *Call.* A sharp 'chruk, chruk' flight note. *Afrikaans.* Grysruglewerik.

517 **Blackeared Finchlark** *Eremopterix australis* L 13 cm
The all-black head and underparts are diagnostic. In flight the male appears all black and therefore might be confused with a widowfinch but closer examination reveals the dark chestnut back. Female differs from female Chestnutbacked and Greybacked finchlarks by being very dark chestnut above and heavily streaked black below, and by lacking the dark belly patch. *Habitat.* Karoo scrub, Kalahari sandveld and cultivated lands. An uncommon and highly nomadic endemic to the central and south-western regions. *Call.* Flight call is a short 'preep' or 'chip-chip'. *Afrikaans.* Swartoorlewerik.

204

511▲   514▼ 515▼   510▲

516▼ 517▼

518 **European Swallow** *Hirundo rustica* L 18 cm
The red throat and forehead, black breast band and deeply forked tail are diagnostic. Distinguished from the similar Angola Swallow by having a black breast band which is more extensive and lacks red below it. *Habitat.* Found in virtually any habitat throughout the region. The most abundant swallow during summer. *Call.* A soft, high-pitched twittering. *Afrikaans.* Europese Swael.

525 **Mosque Swallow** *Hirundo senegalensis* L 24 cm
Much larger than either of the striped swallows and likely to be confused only with the similar-sized Redbreasted Swallow. Distinguished by being much paler in appearance, having white, not buffy wing linings, and a white throat and upper breast. Imm. Redbreasted shows varying degrees of white on throat and breast but always has buffy, not white wing linings, and has the blue on its head extending below the eye. *Habitat.* Open thornveld, often near rivers, and especially where baobab trees occur. Uncommon and localized in the extreme north and north-east. *Call.* A nasal 'harrrrp', as well as a guttural chuckling. *Afrikaans.* Moskeeswael.

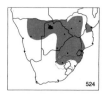

524 **Redbreasted Swallow** *Hirundo semirufa* L 24 cm
This large, very dark swallow can be confused only with the Mosque Swallow from which it differs by having a complete red throat and breast, and dark buffy underwing linings. Imm. has creamy white throat and breast but differs from Mosque Swallow by buffy, not white underwing linings.
Habitat. Open grassy areas in thornveld, and broad-leafed woodland and upland grasslands. Common but thinly distributed in the central, eastern and northern parts of the region. *Call.* A soft warbling song. Twittering notes are uttered in flight. *Afrikaans.* Rooiborsswael.

526 **Greater Striped Swallow** *Hirundo cucullata* L 20 cm
Far larger than Lesser Striped Swallow and appears very much paler, with the striping on buffy underparts discernible only at close range. The orange on the crown and rump is slightly paler and there is no red on the vent.
Habitat. Open grasslands, road culverts, bridges and often found in association with man. Common summer visitor to most parts of the region except the extreme north-east. *Call.* A twittering 'chissick'.
Afrikaans. Grootstreepswael.

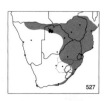

527 **Lesser Striped Swallow** *Hirundo abyssinica* L 16 cm
Smaller and very much darker than Greater Striped Swallow, with heavy black striping on white, not buffy underparts. The rust-coloured rump extends on to the vent in a small patch. *Habitat.* Usually near water, frequently perching in trees or on wires. Common resident and summer visitor to the northern, eastern and south-eastern parts of the region. *Call.* A descending series of squeaky, nasal 'zeh-zeh-zeh-zeh' notes. *Afrikaans.* Kleinstreepswael.

531 **Greyrumped Swallow** *Hirundo griseopyga* L 14 cm
The combination of grey crown and rump contrasting slightly with the blue-black upperparts is diagnostic. In overhead flight could be confused with the House Martin but has a longer, more deeply forked tail. *Habitat.* Over rivers in thornveld, and open grasslands adjacent to lakes and vleis. Uncommon and localized in the north-east and east. *Call.* Flight call note recorded as a 'chraa'. *Afrikaans.* Gryskruisswael.

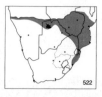

522 **Wiretailed Swallow** *Hirundo smithii* L 13 cm
Similar to but much smaller than Whitethroated Swallow and differs by having an incomplete black breast band and a bright chestnut cap. Much faster in flight than other swallows. *Habitat.* Usually found near water, and especially under bridges. Common in the northern, north-eastern and eastern parts of the region. *Call.* A sharp metallic 'tchik'. *Afrikaans.* Draadstertswael.

206

518▲

524▼   531▲

525▲

526▲   527▼   522▼

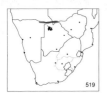

519 Angola Swallow *Hirundo angolensis* L 15 cm
Less agile and slower in flight than the similar European Swallow, it differs
further by having the red on the throat extending on to the breast and being
bordered by a narrow, incomplete black band. Outer tail streamers are much
shorter and the tail less deeply forked than that of the European Swallow.
Habitat. Rivers and bridges, and in association with man. Rare vagrant to the
extreme north-central region. *Call.* Not recorded in our region.
Afrikaans. Angolaswael.

521 Blue Swallow *Hirundo atrocaerulea* L 25 cm
The glossy blue-black plumage and deeply forked tail with extremely long
outer tail feathers are diagnostic. Differs from the superficially similar Black
Saw-wing Swallow by having glossy, not matt black plumage and a much
more deeply forked tail. *Habitat.* Upland grasslands, often bordering forests.
Rare and highly localized in Natal and Zimbabwe. *Call.* A short 'chzzze'
uttered in flight. *Afrikaans.* Blouswael.

523 Pearlbreasted Swallow *Hirundo dimidiata* L 14 cm
Distinguished from Wiretailed Swallow by lacking both the tail streamers and
the breast band, and by having a blue, not chestnut cap. In overhead flight
differs from Greyrumped Swallow and House Martin by having white, not dark
underwings, and a dark rump. *Habitat.* Over bushveld, fresh water and in
association with man. A common but thinly distributed summer visitor to the
south. *Call.* Silent. *Afrikaans.* Pêrelborsswael.

520 Whitethroated Swallow *Hirundo albigularis* L 17 cm
The white throat and black breast band are diagnostic. Sand and Banded
martins also have white throats and dark breast bands but those species are
brown, not glossy blue-black. *Habitat.* Usually near rivers and dams, nesting
under bridges and in culverts. During summer, common throughout the region
except in the north-east and north-west. *Call.* Soft warbles and twitters.
Afrikaans. Witkeelswael.

536 Black Saw-wing Swallow *Psalidoprocne holomelas* L 15 cm
An all-black swallow which differs from the Blue Swallow by having a matt
plumage and a less deeply forked tail with shorter tail streamers.
Distinguished from the Eastern Saw-wing only by its black, not white wing
linings. The smaller size and slow, fluttering flight should eliminate confusion
with any of the black swifts in the region. *Habitat.* Associated with evergreen
forests. Common but localized along the coastal strip from Cape Town to
Moçambique, and inland in the north-east and extreme north. *Call.* A soft
'chrrp' alarm call. *Afrikaans.* Swartsaagvlerkswael.

537 Eastern Saw-wing Swallow *Psalidoprocne orientalis* L 15 cm
Differs from the Black Saw-wing Swallow only by having conspicuous white
wing linings which are easily seen against the black underwings and body.
Habitat. Over evergreen forests, miombo woodland and around rivers in these
areas. Uncommon in the north-east. *Call.* Not recorded in southern Africa.
Afrikaans. Tropiese Saagvlerkswael.

904 Redrumped Swallow *Hirundo daurica* L 18 cm   ☐
Not positively recorded from within our region but it might occur as a vagrant
in the highlands of Zimbabwe or Moçambique.

519▲

521▲

523▲

536▼

537▼

520▲

535 Mascarene Martin *Phedina borbonica* L 13 cm
In the region, the only martin with heavily striped underparts and a white vent. It might be mistaken for the imm. Lesser Striped Swallow but that species has a chestnut crown and rump. At long range the striping on the underparts appears uniform and the bird might seem to resemble the dark form of the Brownthroated Martin, but the large amount of white on its vent should eliminate confusion. *Habitat.* Over thornveld and miombo forest. Rare winter visitor from Madagascar to the extreme north-east. *Call.* Not recorded in southern Africa. *Afrikaans.* Gestreepte Kransswael.

528 South African Cliff Swallow *Hirundo spilodera* L 15 cm
Differs from the striped swallows by having only a slight notch in a square-ended tail, and by having a black breast band and mottling on the upper breast. *Habitat.* Upland grasslands, road culverts and bridges. Common summer visitor. Endemic breeder in the north-western and central regions. *Call.* Indistinct 'chooerp-chooerp' uttered near nesting colonies. *Afrikaans.* Familieswael.

530 House Martin *Delichon urbica* L 14 cm
Swallow-like in appearance, this species is the only martin in the region to have pure white underparts and a white rump. In overhead flight is easily confused with Greyrumped Swallow but has a less deeply forked tail and broader based, shorter wings. *Habitat.* Over dry broad-leafed woodland and thornveld, mountainous terrain, and is coastal in the south. Common but irregular summer visitor. *Call.* A single 'chirrp'. *Afrikaans.* Huisswael.

529 Rock Martin *Hirundo fuligula* L 15 cm
A dark brown martin which might be confused with the all-brown form of the Brownthroated Martin. It differs by being larger, paler on the throat and by having white tail spots which are visible when the tail is slightly spread. *Habitat.* Cliffs, quarries, rocky terrain and human dwellings. Common except in the north-eastern and central areas. *Call.* Soft, indistinct twitterings. *Afrikaans.* Kransswael.

532 Sand Martin *Riparia riparia* L 12 cm
Distinguished from Brownthroated Martin by having a white throat and brown breast band. Differs from the larger Banded Martin by having dark wing linings, a forked tail and by lacking the white eyebrow. *Habitat.* Usually over or near fresh water, roosting in reedbeds in association with European Swallow. Uncommon, localized summer visitor to the north and east. *Call.* Normally silent in Africa. *Afrikaans.* Europese Oewerswael.

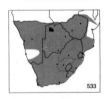

533 Brownthroated Martin *Riparia paludicola* L 12 cm
Occurs in two colour forms, one being dark brown with a small amount of white on the vent, the other, more commonly seen form, having a brown throat and breast with the remainder of the underparts white. Distinguished from Sand Martin by its brown throat. Dark form easily confused with Rock Martin but is smaller, much darker brown below and lacks the white tail spots. *Habitat.* Over freshwater lakes, streams and rivers, especially near sandbanks. Common throughout. *Call.* Soft twittering. *Afrikaans.* Afrikaanse Oewerswael.

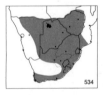

534 Banded Martin *Riparia cincta* L 17 cm
Distinguished from the similar Sand Martin by being much larger, by having white wing linings, a small white eyebrow and a square-ended tail. Imm. resembles ad. but has the upperparts finely scaled with pale buff, and lacks the white eyebrow. Flight is slower and more deliberate than that of the Sand Martin. *Habitat.* In summer frequents upland grasslands; during winter it is more coastal. Small numbers found throughout the region except the dry west. *Call.* Flight call is a 'che-che-che'. Song is a jumble of harsh 'chip-choops' and more melodious warbles. *Afrikaans.* Gebande Oewerswael.

210

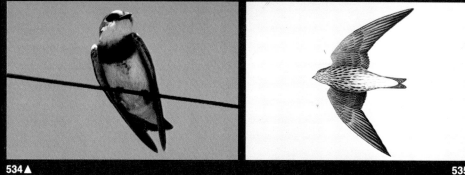

534 ▲

535 ▲

530 ▲

529 ▲   533 ▼ 532 ▼   528 ▲

• 538 **Black Cuckooshrike** *Campephaga flava* L 22 cm
The diagnostic yellow shoulder of the male is sometimes absent, but the orange-yellow gape is a constant characteristic. Differs from Black Flycatcher and Squaretailed Drongo by having a rounded tail, and by its habit of creeping through the tree canopy and not dashing after insects in flight. Female resembles a 'green' cuckoo but is larger, more barred below and has bright yellow outer tail feathers. *Habitat.* Variety of woodlands from coastal bush to evergreen forests, and bushveld. Common in the north, north-east and the coastal strip south to Cape Town. *Call.* A high-pitched, prolonged 'trrrrrrrr'. *Afrikaans.* Swartkatakoeroe.

539 **Whitebreasted Cuckooshrike** *Coracina pectoralis* L 27 cm
Differs from the Grey Cuckooshrike by having a white breast and belly. Female is generally paler than male, sometimes lacking the distinct grey throat and upper breast. Imm. is even paler, with spotted underparts, and upperparts barred with white. *Habitat.* Miombo woodland, riverine forest and thornveld. Uncommon and localized in the north-east. *Call.* Male's call is a 'duid-duid' and female's is a 'tchee-ee-ee-ee'. *Afrikaans.* Witborskatakoeroe.

540 **Grey Cuckooshrike** *Coracina caesia* L 27 cm
The all-grey plumage, white orbital ring, and black patch between the base of the bill and the eye, are diagnostic. Imm. has flight and tail feathers tipped with white and a barred black rump. Female is paler than male and lacks the black patch at bill base. *Habitat.* Evergreen and riverine forests. Uncommon and localized in the south and east, and the highlands of Zimbabwe and Moçambique. *Call.* A soft, thin 'seeeeeeeep'. *Afrikaans.* Bloukatakoeroe.

541 **Forktailed Drongo** *Dicrurus adsimilis* L 25 cm
In our region, the only all-black bird with a deeply forked tail. The Squaretailed Drongo is smaller and has only a slight notch in the tail. Imm. has white scaling on the underparts and forewing and shows a yellow gape. *Habitat.* Diverse. Common throughout except in the central south-west. *Call.* Mimicry and harsh 'tchwaak tchweeek' notes. *Afrikaans.* Mikstertbyvanger.

542 **Squaretailed Drongo** *Dicrurus ludwigii* L 19 cm
Smaller than the Forktailed Drongo and its tail is only slightly notched. Differs from Black Flycatcher (p. 262) by its smaller, rounder head, notched tail, and red, not dark brown eye, which can be seen at close range. Distinguished from male Black Cuckooshrike by notched tail and lack of both yellow shoulder and orange-yellow gape. *Habitat.* Evergreen forest and thick coastal bush. Uncommon and localized in the north-east and east. *Call.* Sharp 'chreep, chreep' and 'cheeeroop' sounds. *Afrikaans.* Kleinbyvanger.

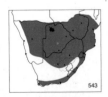

543 **European Golden Oriole** *Oriolus oriolus* L 24 cm
Male differs from male African Golden Oriole by having black wings and less black through the eye. Female is very similar to female African Golden Oriole but is less yellow below, has darker wing coverts and green upperparts. *Habitat.* Thornveld, open broad-leafed woodland and exotic plantations. Common summer visitor to the northern, eastern and south-eastern parts of the region. *Call.* A liquid 'chleeooop'. *Afrikaans.* Europese Wielewaal.

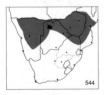

544 **African Golden Oriole** *Oriolus auratus* L 24 cm
A much brighter yellow bird than the European Golden Oriole from which it differs mainly by having yellow wing coverts and a black stripe through and behind the eye. Female distinguished from female European Golden Oriole by being much yellower and having yellowish green, not black wing coverts. *Habitat.* Thornveld, riverine forest and miombo woodland. Uncommon and localized in the north. *Call.* A liquid whistle 'fee-yoo-fee-yoo'. *Afrikaans.* Afrikaanse Wielewaal.

542 ▲

540 ▲

543 ▲

538 ▼   539 ▲

541 ▼

544 ▼

545 Blackheaded Oriole *Oriolus larvatus* L 24 cm
The black head is diagnostic in all plumages. If seen at long range might be confused with the rare Greenheaded Oriole but has a yellow, not green back. Imm. lacks coral red bill of ad. and its head is dark brown, flecked with green. *Habitat.* Frequents a wide range of woodland and forest habitats and has adapted well to exotic plantations. *Call.* A clear, liquid whistle 'pooodleeoo' and a harsher 'kweeer' note. *Afrikaans.* Swartkopwielewaal.

546 Greenheaded Oriole *Oriolus chlorocephalus* L 24 cm
The green head, yellow collar and green back are diagnostic. If seen at long range the head appears black which might lead to confusion with the Blackheaded Oriole, but the back also appears black, not yellow as in the Blackheaded. Imm. has dull yellow underparts, breast slightly streaked with olive and has a pale olive wash on head and throat. *Habitat.* Evergreen montane forests. Rare and, in our region, found only on Mt Gorongosa in Moçambique. *Call.* The liquid call typical of orioles. *Afrikaans.* Groenkopwielewaal.

547 Black Crow *Corvus capensis* L 50 cm
Larger than the House Crow and differs by being entirely glossy black with a long, slender, slightly decurved bill. Imm. is generally duller, lacking the glossy blue-black plumage of ad. Usually seen in pairs but does form small flocks. *Habitat.* Upland grasslands, open country, cultivated fields and dry desert regions. Common throughout but absent from the central and north-eastern regions. *Call.* Normal crow-like 'kah-kah' and other deep bubbling notes. *Afrikaans.* Swartkraai.

549 House Crow *Corvus splendens* L 42 cm
The only grey-bodied crow in the region. The head, wings and tail are a glossy blue-black. Imm. has body colour greyish brown, but in other respects it resembles the ad. *Habitat.* Not often seen away from human habitation, as it scavenges food scraps from gardens, drive-in cinemas and restaurants. A small population has recently established itself in Durban. Records also exist from Moçambique and the Cape. *Call.* A higher pitched 'kaah, kaah' than other crows. *Afrikaans.* Huiskraai.

548 Pied Crow *Corvus albus* L 50 cm
The only white-bellied crow in the region. At long range distinguished from Whitenecked Raven by longer tail and smaller head. A common city dweller, sometimes forming roosts of several hundred individuals. *Habitat.* Occurs in virtually every habitat. Common, conspicuous bird found throughout the region, but absent from the central region. *Call.* Typical 'kwaaa' or 'kwooork' cawing. *Afrikaans.* Witborskraai.

550 Whitenecked Raven *Corvus albicollis* L 54 cm
The massive black bill with a white tip, and the white crescent on the hind neck are diagnostic. In flight and at long range, distinguished from Black and Pied crows by its much broader wings, larger, heavier head and short, broad tail. *Habitat.* Strictly a species of mountainous terrain but may be seen over upland grasslands and open country. Common in suitable areas in the south, east and north-east. *Call.* A deep, throaty 'kwook'. *Afrikaans.* Withalskraai.

548▲   550▼   547▲

545▼   546▼   549▼

551 Southern Grey Tit *Parus afer* L 13 cm
Distinguished from the Ashy Tit and Northern Grey Tit by having a distinctive brownish grey, not blue-grey back, buffy belly and flanks, and a proportionately shorter tail. Overall appearance is of a dowdier, less clean-cut species than either the Ashy or Northern Grey Tit. *Habitat.* Fynbos and Karoo scrub. A common endemic to the southern Cape. *Call.* Song is a sharp 'klee-klee-klee-cheree-cheree'. Alarm note is a harsh 'chrrr'. *Afrikaans.* Piet-tjou-tjougrysmees.

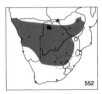

552 Ashy Tit *Parus cinerascens* L 13 cm
Differs from Southern Grey Tit by having a blue-grey, not brown-grey back, and blue-grey, not buffy flanks and belly. Generally much darker than Northern Grey Tit, with less white in the wings. *Habitat.* Thornveld, dry broad-leafed woodland and riverine scrub. Common endemic to the central and north-western regions. *Call.* Indistinguishable from that of Southern Grey Tit. *Afrikaans.* Acaciagrysmees.

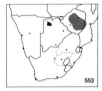

553 Northern Grey Tit (Miombo Grey Tit) *Parus griseiventris* L 13 cm
Distinguished from the Southern Grey Tit by having a blue-grey, not brownish grey back and pale blue-grey, not buffy flanks. Much paler in appearance than the Ashy Tit with a greater expanse of white in the wings.
Habitat. Confined to miombo woodland. Common in the north-east of the region. *Call.* Not noticeably different from that of either the Ashy Tit or Southern Grey Tit. *Afrikaans.* Miombogrysmees.

554 Southern Black Tit *Parus niger* L 16 cm
Male has much less white in the wings than Carp's Tit and has grey barred undertail coverts. Female differs from female Carp's Tit by being paler grey below and having less white in the wings. *Habitat.* Variety of woodlands, from evergreen forests to thornveld. Common in the north, east and south-east. *Call.* A harsh, chattering 'chrr-chrr-chrr' and a musical 'phee-cher-phee-cher'. *Afrikaans.* Gewone Swartmees.

555 Carp's Tit *Parus carpi* L 14 cm
A smaller bird than the Southern Black Tit from which it differs by having almost totally white wings and by lacking the grey barred undertail coverts. Female differs from female Southern Black Tit by being much darker below and by having more white in the wings. *Habitat.* Dry thornveld, broad-leafed woodland and dry riverine forest. Common endemic to northern Namibia. *Call.* Very similar to that of the Southern Black Tit. *Afrikaans.* Ovamboswartmees.

556 Rufousbellied Tit *Parus rufiventris* L 15 cm
Unlikely to be confused with any other tit in the region. The dark head and breast, lack of white cheek patches, together with its rufous belly, flanks and vent, are diagnostic. Its bright yellow eye is conspicuous at close range. Imm. is duller, has a brown eye and buffish edges to its wing feathers.
Habitat. Found mainly in miombo woodland. Uncommon in the north-east and extreme north-west. *Call.* A harsh tit-like 'chrrr chrrr' and a clear 'chick-wee, chick-wee' song. *Afrikaans.* Swartkopmees.

559 Spotted Creeper *Salpornis spilonotus* L 15 cm
Much smaller than any woodpecker in the region. The brown upperparts, heavily spotted with white, and the thin, decurved bill are diagnostic. Creeps jerkily up tree trunks. *Habitat.* Confined to miombo woodland. Uncommon and thinly distributed in the north-east. *Call.* A fast, thin 'sweepy-swip-swip-swip' and a harsher 'keck-keck'. *Afrikaans.* Boomkruiper.

551 ▲

552 ▲

556 ▼ 554 ▲

553 ▲ 555 ▼

559 ▼

560 Arrowmarked Babbler *Turdoides jardineii* L 24 cm
When flying, this species can be distinguished from Hartlaub's Babbler by its dark rump. Seen at close range, it will be noticed that the head and breast are streaked, not scalloped with white as in Hartlaub's Babbler. *Habitat.* Thornveld, dry broad-leafed woodland and stands of eucalyptus. Common in the northern, north-eastern and eastern parts of the region. *Call.* Small parties utter a harsh churring and cackling. *Afrikaans.* Pylvlekkatlagter.

562 Hartlaub's Babbler *Turdoides hartlaubii* L 26 cm
Superficially resembles the Arrowmarked Babbler but has a diagnostic white rump and vent, and has white scalloping, not streaking on its head and body. Imm. is similar to ad. but is much paler, especially on the throat and breast. *Habitat.* Parties occur in reedbeds and surrounding woodland. Common in the extreme north-central region. *Call.* Extremely vocal, with a loud 'kwekkwekkwek' or 'papapapapapa' call. *Afrikaans.* Witkruiskatlagter.

561 Blackfaced Babbler *Turdoides melanops* L 28 cm
In coloration, a more uniform grey-brown than the Arrowmarked and Hartlaub's babblers, with only faint white streaking on the head. A small black patch at the base of the bill, and the bright yellow eye are diagnostic. It is less demonstrative, more furtive than other babblers. *Habitat.* Stretches of tall grass in open broad-leafed woodland. Uncommon and confined to the extreme north-west. *Call.* A nasal 'wha-wha-wha' and a harsh, fast 'papapapa'. *Afrikaans.* Swartwangkatlagter.

563 Pied Babbler *Turdoides bicolor* L 26 cm
In the region, the only babbler with a white head, back and underparts. Imm. initially has a pale brown head and body but it whitens with age. Conspicuous, as small parties flit from bush to bush. *Habitat.* Thornveld, dry broad-leafed woodland and dry riverine forest. Common endemic to the central and dry north-western regions. *Call.* 'Kwee kwee kwee kweer', a higher pitched babbling than the other species. *Afrikaans.* Witkatlagter.

564 Barecheeked Babbler *Turdoides gymnogenys* L 24 cm
Easily distinguished from the Pied Babbler by its dark back and by the lack of white on its wing coverts. Has a small area of bare black skin below and behind the eye. Imm. could be confused with the imm. Pied Babbler but is generally much darker, especially on the back and nape, and is usually seen in the company of ads. of its own species. *Habitat.* Dry broad-leafed woodland, frequenting rivercourses and wooded koppies. Uncommon and thinly distributed endemic to the north-west. *Call.* Typical babbler 'kerrrakerrra-kek-kek-kek'. *Afrikaans.* Kaalwangkatlagter.

610 Boulder Chat *Pinarornis plumosus* L 25 cm
The brownish black plumage and the white-tipped outer tail feathers are diagnostic. In flight, a row of small white spots on the edge of the primary and secondary coverts can be seen. Runs and bounds over large boulders, occasionally raising its tail well over its back when landing. *Habitat.* Well-wooded terrain with large granite boulders. Locally common in and virtually confined to suitable areas in Zimbabwe. *Call.* A clear, sharp whistle and softer 'wink, wink' call. *Afrikaans.* Swartberglyster.

560 ▲   561 ▼ 563 ▼   562 ▲

                              564 ▼ 610 ▼

566 Cape Bulbul *Pycnonotus capensis* L 21 cm
Overall much darker in appearance than the Redeyed and Blackeyed bulbuls, with the dark brown on the underparts extending on to the lower belly. The white eye-ring is diagnostic. Imm. lacks the white eye-ring but is, like the ad., significantly darker below than the Blackeyed and Redeyed bulbuls. *Habitat.* Fynbos, coastal scrub, riverine forests and exotic plantations. Common endemic to the southern Cape. *Call.* A sharp, liquid whistle: 'peet-peet-patata'. *Afrikaans.* Kaapse Tiptol.

567 Redeyed Bulbul *Pycnonotus nigricans* L 21 cm
The red eye-ring is diagnostic and distinguishes this species from the Blackeyed and Cape bulbuls. Head colour is darker than that of the Blackeyed and it contrasts with the greyish buff collar and breast. Imm. can be differentiated from imm. Blackeyed Bulbul by its pale pink eye-ring. *Habitat.* Rivercourses in dry thornveld. A common endemic, replacing the Blackeyed Bulbul in the west. *Call.* Liquid whistles similar to those of the Blackeyed Bulbul. *Afrikaans.* Rooioogtiptol.

568 Blackeyed Bulbul *Pycnonotus barbatus* L 22 cm
Lacks the red eye-ring of the Redeyed Bulbul and further differs by having less contrast between dark head and body, and by lacking the buffy grey collar. *Habitat.* Frequents a wide variety of habitats, from thornveld to evergreen forests and parks and gardens. An obvious and abundant bird in the east and north-east. *Call.* A harsh, sharp 'kwit, kwit, kwit' given when alarmed or when going to roost. Song is a liquid 'cheloop chreep choop'. *Afrikaans.* Swartoogtiptol.

565 Bush Blackcap *Lioptilus nigricapillus* L 17 cm
A small, unmistakable bulbul with a black cap and bright orange-pink bill and legs. Shy and unobtrusive, it is usually found in pairs in dense foliage. *Habitat.* Thick evergreen forest and scrub bush in mountain river valleys. Uncommon endemic to the south-eastern and eastern highlands. *Call.* Very similar to that of Blackeyed Bulbul but more varied and melodious. *Afrikaans.* Rooibektiptol.

575 Yellowspotted Nicator *Nicator gularis* L 23 cm
The bright yellow spots on the wing coverts are diagnostic of this greenish bird. Could be confused with the female Black Cuckooshrike but that species is heavily barred below. In behaviour acts like a bush shrike, moving slowly through thick tangles, and is very difficult to see. *Habitat.* Dense riverine and coastal forest and scrub, particularly on sandy soil. Common, but easily overlooked, in the east, north-east and extreme north. *Call.* A short, rich, throaty chortle. *Afrikaans.* Geelvleknikator.

566▲ 567▲

568▲ 565▼ 575▼

569 Terrestrial Bulbul *Phyllastrephus terrestris* L 21 cm
A dull, earth-brown bulbul with greyish underparts. The white throat contrasting with the dark head is the best field character. In its search for food, it shuffles around on the forest floor, scattering earth and leaf litter with its feet and bill. *Habitat.* Thick thornveld, and evergreen and riverine forests. Common in certain areas along the southern and eastern coastal regions, and inland in the north-east and north. *Call.* A soft, chattering 'trrup cherrup trrup' given by small foraging groups. *Afrikaans.* Boskrapper.

570 Yellowstreaked Bulbul *Phyllastrephus flavostriatus* L 20 cm
Resembles the Sombre Bulbul but has a red, not white eye, a much longer, thinner bill, and indistinct yellow streaking on its underparts. Best identified by its unusual woodpecker-like action of creeping about tree trunks, continually flicking its wings over its back. *Habitat.* Mid-stratum and canopy of moist evergreen forests. Common in suitable habitat in the east and north-east of the region. *Call.* 'Weet-weet-weet' and a sharp 'kleeet-kleeat'. *Afrikaans.* Geelstreepboskruiper (-tiptol).

571 Slender Bulbul *Phyllastrephus debilis* L 14 cm
The smallest green bulbul of the region. Found in association with the Sombre Bulbul, which it resembles, but is much smaller, paler below, and has a greyish crown and cheeks. *Habitat.* Thick coastal forest and miombo woodland. Common only in the north-east. *Call.* A shrill 'shriiip' and a bubbling song. *Afrikaans.* Kleinboskruiper (-tiptol).

572 Sombre Bulbul *Andropadus importunus* L 23 cm
Most likely to be confused with the Yellowbellied Bulbul: this is particularly true of the north-eastern race *hypoxanthus* which is very yellow below, but it can be distinguished by its white, not red eye, and greenish, not brown crown. The other two races are drab green, and lack any diagnostic field characters except the white eye. Difficult to locate in a leafy tree canopy.
Habitat. Evergreen forests and coastal bush. Common to abundant in suitable habitat, from Cape Town eastwards along the coast, and inland further north. *Call.* Normal call is a piercing 'weeewee', followed by a liquid chortle. *Afrikaans.* Gewone Willie.

573 Stripecheeked Bulbul *Andropadus milanjensis* L 21 cm
Distinguished from the similar Sombre Bulbul by having a darker head with a pale eye-ring, a dark, not white eye and indistinct white striping on the cheeks. A shy forest resident, it is difficult to locate in a tree canopy. *Habitat.* Evergreen forests and their edges. Locally very common in the north-east. *Call.* A throaty 'chrrup-chip-chrup-chrup'. *Afrikaans.* Streepwangwillie (-tiptol).

574 Yellowbellied Bulbul *Chlorocichla flaviventris* L 23 cm
Distinguished from the Sombre Bulbul by having yellow underparts and a red, not white eye. The race *hypoxanthus* of the Sombre Bulbul also has a very yellow breast and belly but differs by having a greenish, not brown crown, and a white eye. Not often found in canopy, preferring the thick tangles at a lower level. *Habitat.* Evergreen and riverine forests, thornveld and coastal scrub. Common in the east, north-east and extreme north. *Call.* A monotonous, nasal 'neh-neh-neh-neh'. *Afrikaans.* Geelborswillie (-tiptol).

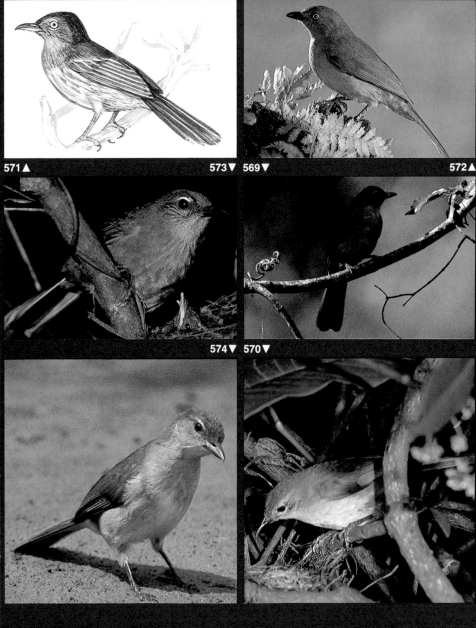

571▲   573▼ 569▼   572▲

574▼ 570▼

576 Kurrichane Thrush *Turdus libonyana* L 22 cm
The white, faintly speckled throat, broad black malar stripes and bright orange bill are diagnostic. The Olive Thrush also has a bright orange-yellow bill but it has a dark throat and lacks the malar stripes. *Habitat.* Thornveld, open woodland, and parks and gardens in drier regions. Common in the east, north-east and north. *Call.* A loud, whistling 'peet-peeoo'. *Afrikaans.* Rooibeklyster.

577 Olive Thrush *Turdus olivaceus* L 24 cm
Distinguished from the Kurrichane Thrush by being darker in appearance, by having a yellowish, not bright orange bill, a dark-speckled throat, and by lacking the black malar stripes. Some birds show very bright orange underparts, in which case they are then distinguishable from the Orange Thrush by the lack of white bars on the wing coverts. *Habitat.* Moist evergreen forests, coastal scrub, and has adapted to parks and gardens. *Call.* A sharp 'chink' or thin 'tseeep'. Song is a fluty 'wheeet-tooo-wheeet'.
Afrikaans. Olyflyster.

579 Orange Thrush *Turdus gurneyi* L 23 cm
Differs from both the Olive and Kurrichane thrushes by the dark bill, bright orange throat and breast, and conspicuous white stripes on the wing coverts. Imm. is spotted with buff on the upperparts, and mottled black and orange below, but it still shows the diagnostic white stripes on the wing coverts. *Habitat.* Coastal and inland evergreen forests. A shy, uncommon bird found in isolated localities in the east. *Call.* Typical thrush-like 'tseeep'. Song consists of a clear melodious 'chee-cheeleeroo-chruup' and many other phrases.
Afrikaans. Oranjelyster.

580 Groundscraper Thrush *Turdus litsitsirupa* L 22 cm
Distinguished from the Spotted Thrush by lacking white bars on its wing coverts and by having bolder, more contrasting black face markings. In flight shows a chestnut panel in the wing. Differs from the much smaller Dusky Lark (p. 198) by having bolder face markings, grey legs, and is rarely seen in flocks. *Habitat.* Dry thornveld, open broad-leafed woodland, and parks and gardens. Common in the northern, central and eastern parts of the region. *Call.* This bird's specific name is onomatopoeic of its song: a loud, clear and varied whistling phrase. *Afrikaans.* Gevlekte Lyster.

578 Spotted Thrush *Turdus fischeri* L 23 cm
Distinguished from the Groundscraper Thrush by having bold white stripes on the wing coverts and by having flesh-coloured, not grey legs. This and the Groundscraper Thrush superficially resemble the Dusky Lark but both are larger, show paler backs and are unlikely to be found in flocks.
Habitat. Rank growth along rivers in coastal forests. Uncommon in the south-east. *Call.* A soft, thin 'tseeeep'. Song is more robin-like than that of other thrushes. *Afrikaans.* Natallyster.

579 ▲

577 ▼

576 ▼ 578 ▼

580 ▲

581 Cape Rock Thrush *Monticola rupestris* L 21 cm
The largest rock thrush in the region. The male differs from the male Sentinel Rock Thrush by having a brown, not blue back and forewing, and by the blue on its throat not extending on to the breast region. The female has much richer red underparts than other female rock thrushes. *Habitat.* Mountainous and rocky terrain, both at the coast and inland. Common endemic to the south and east. *Call.* Song is a 'tsee-tsee-tseet-chee-chweeeoo' whistling. *Afrikaans.* Kaapse Kliplyster.

582 Sentinel Rock Thrush *Monticola explorator* L 18 cm
Male distinguished from male Cape Rock Thrush by having a blue, not brown back and forewing, with the blue extending well on to the breast. Could be confused with the Short-toed Rock Thrush (although their ranges do not overlap), but this species has a much darker blue forehead and crown. Female differs from the female Cape Rock Thrush by being smaller, and having a lighter, mottled, not uniform brown breast. *Habitat.* Mountainous and rocky terrain, often found in association with the Cape Rock Thrush. A locally common endemic to the south and east. *Call.* A whistle very similar to that of the Cape Rock Thrush. *Afrikaans.* Langtoonkliplyster.

583 Short-toed Rock Thrush *Monticola brevipes* L 18 cm
In the male, the very pale blue forehead, crown and nape are diagnostic: in some birds they are so pale that they appear white. Some individuals show a darker blue head but they usually retain a pale blue eyebrow stripe. The female has an extensively striped throat and breast. *Habitat.* Thickly wooded koppies and rocky slopes. Common endemic to the dry central and western regions. *Call.* A thin 'tseeep'. *Afrikaans.* Korttoonkliplyster.

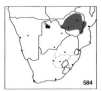

584 Miombo Rock Thrush *Monticola angolensis* L 18 cm
The blue-grey crown and back of the male are mottled with black, thus eliminating confusion with any other rock thrush. The female's distinctly mottled upperparts and bold malar stripes separate this species from other female rock thrushes. *Habitat.* Uncommon and restricted to miombo woodland in the north-east. *Call.* A high-pitched warbling song. *Afrikaans.* Angolakliplyster.

611 Cape Rockjumper *Chaetops frenatus* L 25 cm
Similar in almost every respect to Orangebreasted Rockjumper but is slightly larger and has a dark rufous belly and rump. Female and imm. are darker buff below than the female Orangebreasted Rockjumper and have a more boldly marked head pattern. *Habitat.* Rocky mountain slopes and scree interspersed with grass. Common but localized in the southern Cape. *Call.* A clear 'wheeoo' whistle and a warbling song. *Afrikaans.* Kaapse Berglyster.

612 Orangebreasted Rockjumper *Chaetops aurantius* L 21 cm
Differs from the Cape Rockjumper only by having a much paler belly, flanks and rump and by being slightly smaller. As their distribution ranges are mutually exclusive, confusion between these species is unlikely.
Habitat. Usually found above an altitude of 2 000 m on rocky mountain slopes and grassy hillsides scattered with boulders. Endemic to the high mountain ranges of Lesotho, Natal and the north-eastern Cape. *Call.* A sharp 'chreee-chreee-chreee' whistle and a harsher, grating 'crrree-crrree' call. *Afrikaans.* Oranjeborsberglyster.

581 ▲   583 ▼ 584 ▼   582 ▲

611 ▼ 612 ▼

588 Buffstreaked Chat *Oenanthe bifasciata* L 17 cm
Distinguished from Capped Wheatear by having conspicuous buffy white scapulars and a black, not white throat. Female differs from imm. Capped Wheatear by having a buff, not white rump. Imm. is mottled with black and buff on its upper- and underparts and has a rust-coloured rump.
Habitat. Mountainous and rocky terrain. A common but localized endemic to the central and south-eastern regions. *Call.* Loud, rich warbling, including mimicry of other birds' songs. *Afrikaans.* Bergklipwagter.

587 Capped Wheatear *Oenanthe pileata* L 18 cm
The ad., with its conspicuous white forehead and eyebrow, and black cap, is unmistakable. Imm. is distinguished from the similar European Wheatear by its larger size and by having only a small amount of white at the base of its outer tail feathers. Imm. could also be confused with the female Buffstreaked Chat but its white, not buff rump should aid identification. *Habitat.* Barren sandy or stony areas and short grasslands. Common but thinly distributed throughout, except in the eastern lowlands. *Call.* A 'chik-chik' alarm note. Song is a loud warbling with slurred chattering. *Afrikaans.* Hoëveldskaapwagter.

585 European Wheatear *Oenanthe oenanthe* L 15 cm
Male in breeding plumage is unmistakable with its blue-grey crown and back, and black ear patches. Females and non-breeding males could be confused with imm. Capped Wheatear but show white rumps and sides to their tails, lending a 'T' pattern to the black area of the tail. *Habitat.* Dry stony or sandy areas with stunted scrub growth. Rare summer visitor from Europe and Asia to the extreme north of the region. *Call.* Harsh 'chak-chak'. Song not recorded in southern Africa. *Afrikaans.* Europese Skaapwagter.

589 Familiar Chat *Cercomela familiaris* L 15 cm
Differs from Tractrac and Sicklewinged chats by being a darker grey-brown below, with a richer chestnut rump and outer tail feathers. Like other chats it flicks its wings when at rest, but this species also 'trembles' its tail.
Habitat. Rocky and mountainous terrain. Common, but absent from the Okavango region and north-eastern desert areas. *Call.* A soft 'shek-shek'. *Afrikaans.* Gewone Spekvreter.

596 Stonechat *Saxicola torquata* L 14 cm
The male is unmistakable with its black head, white on neck, wings and rump, and its rufous breast. Female is duller but retains the white in the wings and shows a buff rump. In all plumages, both sexes of this species are distinguishable from the Whinchat, as neither ever shows a white eyebrow stripe. *Habitat.* Upland grasslands and open, treeless areas with short scrub. Common throughout but absent from the drier western region. *Call.* A 'weet-weet' followed by a harsh 'chak'. *Afrikaans.* Gewone Bontrokkie.

597 Whinchat *Saxicola rubetra* L 14 cm
Differentiated from the female Stonechat by having a well-defined buffy eyebrow stripe, a more extensive white wing bar on its coverts, a white patch at the base of its primaries, and white sides to the base of its tail. *Habitat.* Open grasslands with patches of stunted scrub. Rare summer visitor to the extreme north, with scattered records further south. *Call.* Not recorded in southern Africa. *Afrikaans.* Europese Bontrokkie.

907 Pied Wheatear *Oenanthe pleschanka* L 15 cm
Superficially resembles the Mountain Chat but is smaller, has a black face and throat, clear white underparts and lacks the white shoulder patches. Female indistinguishable from female European Wheatear. *Habitat.* Usually dry stony regions with scattered scrub. Rare vagrant, with one record from Natal. *Call.* Not recorded in our region.

588 ▲

587 ▲   585 ▼

589 ▼

907 ▼   596 ◢

597 ▼

586 Mountain Chat *Oenanthe monticola* L 20 cm
Males show variable plumage coloration, but all have a white rump, white sides to the tail and a white shoulder patch. Differs from male Arnot's Chat by having a white, not black rump. The nondescript female is very similar to the Karoo Chat but has a white rump and the basal half of the outer tail feathers white. *Habitat.* Mountainous and rocky terrain. Common endemic to the southern, central and western parts of the region. *Call.* A clear, thrush-like whistling song, interspersed with harsh chatters. *Afrikaans.* Bergwagter.

594 Arnot's Chat *Thamnolaea arnoti* L 18 cm
Distinguished from the similar Mountain Chat by having a black rump and base of tail, and black outer tail feathers. Female lacks the white cap but has a conspicuous white throat and upper breast. *Habitat.* Miombo and mopane woodland. Common in the north-east and extreme northern parts of the region. *Call.* A quiet, whistled 'fick' or 'feee'. *Afrikaans.* Bontpiek.

595 Anteating Chat *Myrmecocichla formicivora* L 18 cm
In flight, easily distinguished from other chats by the conspicuous white patches on the primaries. At rest appears very plump and short-tailed, with an upright stance. The plumage ranges from a dark brown to a mottled black, with males showing a small white shoulder patch. *Habitat.* Upland grasslands dotted with termite mounds, and open, sandy or stony areas. A common, widespread endemic but absent from the north-east and most coastal regions. *Call.* A short, sharp 'peek' or 'piek'. *Afrikaans.* Swartpiek.

593 Mocking Chat *Thamnolaea cinnamomeiventris* L 23 cm
The male's black plumage, bright chestnut belly, vent and rump, and the white shoulder patch are diagnostic. Female is very different in appearance: dark grey replaces the black, and she has a red-brown belly, vent and rump. *Habitat.* Bases of cliffs and thickly wooded rocky slopes. Common but localized in the south-eastern, eastern and northern regions. *Call.* Song is a loud, melodious mixture of mimicked bird song. *Afrikaans.* Dassievoël.

590 Tractrac Chat *Cercomela tractrac* L 15 cm
Smaller than the Karoo Chat and lacks the all-white outer tail feathers of that species. Paler and greyer than both the Familiar and Sicklewinged chats, it shows a white or pale buff rump. Namib desert race is almost white, with darker wings and tail. *Habitat.* Karoo and desert scrub. Common endemic to the dry west. *Call.* A soft, fast 'tactac'. Song is a quiet musical bubbling. *Afrikaans.* Woestynspekvreter.

591 Sicklewinged Chat *Cercomela sinuata* L 15 cm
The dark upperparts contrasting with paler underparts distinguish this species from the more uniformly coloured Familiar Chat. It further differs by having the paler chestnut on the rump confined to the tail base, and by its longer legged, more upright stance. In comparison with the Familiar Chat, spends more time on the ground and can run swiftly. *Habitat.* Short grasslands and barren, sandy or stony areas. A common but thinly distributed endemic to the south. *Call.* A very soft, typically chat-like 'chak-chak'. *Afrikaans.* Vlaktespekvreter.

592 Karoo Chat *Cercomela schlegelii* L 18 cm
Generally paler than the similar grey phase female Mountain Chat, it also differs by always having a grey, not white rump, and completely white outer tail feathers. Its larger size eliminates confusion with the Tractrac and Sicklewinged chats. *Habitat.* Karoo and desert scrub. A common endemic to the central and dry western regions. *Call.* A typically chat-like 'chak-chak' or 'trrat-trrat'. *Afrikaans.* Karoospekvreter.

594 ▲   595 ▼

592 ▲   590 ▼

591 ▲   586 ▼   593 ▼

598 Chorister Robin *Cossypha dichroa* L 20 cm
The dark upperparts, yellow-orange underparts, and lack of white eye-stripe render this large robin distinctive. Superficially resembles the Natal Robin but lacks the powder blue forewings and orange face of that species. *Habitat.* Inland and coastal evergreen forests. A common but localized endemic to the east and south-east. *Call.* A loud, clear song, comprising mimicry of other forest birds. *Afrikaans.* Lawaaimakerjanfrederik.

600 Natal Robin *Cossypha natalensis* L 18 cm
Distinguished from all other robins by having a powder blue back, crown and nape, and bright red-orange face and underparts. Imm. mottled above and below with buff and black, but shows a red tail with a dark centre. *Habitat.* Dense thickets and tangles in evergreen forests. Common in the south-eastern, eastern and northern parts of the region. *Call.* Contact sound is a soft 'seee-saw, seee-saw'. Song is a rambling series of melodious phrases, and mimicry. *Afrikaans.* Nataljanfrederik.

599 Heuglin's Robin *Cossypha heuglini* L 20 cm
Distinguished from the similarly sized and coloured Chorister Robin by having a paler back and a broad, conspicuous white eyebrow stripe. Imm. duller than ad. with heavy buff and brown spotting on upper- and underparts. *Habitat.* Dense riverine thickets and tangles, and gardens and parks. Common in the north-east and north. *Call.* A loud, explosive song with repeated, varied phrases. *Afrikaans.* Heuglinse Janfrederik.

601 Cape Robin *Cossypha caffra* L 18 cm
Best distinguished from Heuglin's Robin by its smaller size, shorter white eyebrow stripe, and by having the orange on the underparts confined to the throat and upper breast: the rest of the underparts are washed with grey. Imm. is brownish, heavily mottled with buff and black, and has a red tail with a dark centre. *Habitat.* Montane river valley bush and scrub, evergreen forests, and is widely dispersed in fynbos. Common in the southern sector of the region and in the Eastern Highlands of Zimbabwe. *Call.* Alarm call is a harsh 'chrrrrr'. Song is a soft, melodious 'cherooo-weet-weet-weeeet' phrase. *Afrikaans.* Gewone Janfrederik.

602 Whitethroated Robin *Cossypha humeralis* L 17 cm
The only robin in the region to have white scapulars and a white throat and breast. The rufous tail also distinguishes this species from other robins as it has a black tip to the dark centre. Usually seen as it dashes for cover, it gives the impression of a small black and white bird with a red tail. *Habitat.* Dry thornveld, thickets and riverine scrub. Common endemic to the northern, central and eastern districts. *Call.* A repeated 'seet-cher-seet-cher' whistled phrase. *Afrikaans.* Witkeeljanfrederik.

606 Starred Robin *Pogonocichla stellata* L 16 cm
This small, dark blue and yellow robin is unmistakable. The white 'stars' on the throat and forehead are difficult to see in the field. As the bird flits through the forest undergrowth, the orange-yellow patches on the tail are conspicuous. Imm. is greenish above and completely yellow below but has the same tail pattern as the ad. *Habitat.* Coastal and inland evergreen forests. Common, but localized and easily overlooked, in suitable habitat in the east. *Call.* A soft 'chuk-chuk' or 'zit-zit' is given. Song is a quiet, melodious warbling. *Afrikaans.* Witkoljanfrederik.

232

598 ▲   599 ▼ 601 ▼   600 ▲

602 imm ▼ 606 ▼

616 Brown Robin *Erythropygia signata* L 18 cm
Most likely to be confused with the Bearded Robin in the north of its range. Differs by lacking the rust-coloured flanks of that species, and has less bold white eyebrow and malar stripes. A shy, skulking bird which enters open forest glades at dawn and dusk. *Habitat.* The thick tangles of coastal and evergreen forests. An uncommon and thinly distributed endemic to the east and south-east. *Call.* A melodious 'twee-choo-sree-sree' introduces a varied song. Alarm note is a soft 'krrrrr'. *Afrikaans.* Bruinwipstert.

617 Bearded Robin *Erythropygia quadrivirgata* L 18 cm
The broad white eyebrow stripe finely edged with black, black malar stripes, and rust-coloured flanks, breast and rump are diagnostic. Differs from the duller Brown Robin by having bolder head markings and by the rust-coloured flanks. *Habitat.* Prefers drier broad-leafed woodland than the Brown Robin and is often found in thornveld. Common, but easily overlooked, in the north and north-east. *Call.* A clear, penetrating song of often-repeated mixed phrases. *Afrikaans.* Baardwipstert.

614 Karoo Robin *Erythropygia coryphaeus* L 17 cm
Lacks the breast streaking of the Whitebrowed Robin and is much darker above and below than the Kalahari Robin. Distinctive features are the dull grey-brown plumage, small white eyebrow stripe, and white tips to the dark tail. *Habitat.* Karoo scrub and fynbos. A common endemic to the south-western parts of the region. *Call.* A harsh, chittering 'tchik, tchik, tcheet'. *Afrikaans.* Slangverklikker.

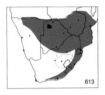

613 Whitebrowed Robin (Whitebrowed Scrub Robin)
Erythropygia leucophrys L 15 cm
Resembles the Kalahari Robin but differs by having a heavily streaked, not unmarked breast, and white bars on the wing coverts. The breast streaking eliminates confusion with Brown and Bearded robins. *Habitat.* A wide range of both wet and dry broad-leafed woodlands and thornveld. Common in the north and east; is replaced by the Kalahari Robin in the west. *Call.* A harsh 'trrrrrr' alarm note and a very varied fluty song. *Afrikaans.* Gestreepte Wipstert.

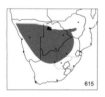

615 Kalahari Robin *Erythropygia paena* L 17 cm
The conspicuous white sides to the tip of the tail contrast with the remainder of the black tail and render this robin unmistakable. The rest of the plumage is buff to greyish buff, with the rump and base of the tail being bright russet. *Habitat.* Dry thornveld, thicket and tangled growth around waterholes and dry riverbeds. Common endemic to the dry west and north-west. *Call.* A varied whistling song interspersed with harsh notes. *Afrikaans.* Kalahariwipstert.

234

616▲   614▲

613▲   617▼ 615▼

608 Gunning's Robin *Sheppardia gunningi* L 14 cm
A small, dull forest robin, brown above and orange below. The diagnostic field characters are difficult to see: powder blue forewings and a small white eyebrow stripe. A furtive, infrequently seen robin. *Habitat.* Thick understorey of evergreen forests. Common in the north-east. *Call.* Not recorded in southern Africa. *Afrikaans.* Gunningse Janfrederik.

607 Swynnerton's Robin *Swynnertonia swynnertoni* L 14 cm
Distinguished from the similar Starred Robin by having an orange, not yellow breast, and a diagnostic black and white breast band. It also has an all-black, not an orange-yellow and black tail. Imm. is a mottled buff-brown with a white belly and flanks. *Habitat.* Evergreen montane forests and ravines. Uncommon endemic to the highlands of Zimbabwe and Moçambique. Recently found north of our region. *Call.* A soft 'si-si' is uttered. *Afrikaans.* Bandkeeljanfrederik.

605 Whitebreasted Alethe *Alethe fuelleborni* L 20 cm
The only robin-like bird in the region to have completely white underparts. The upperparts and the short tail are a rich russet colour. A robust, plump bird which hops about the forest floor. *Habitat.* The thicker tangles of moist, evergreen forests. Rare, found only in the north-east. *Call.* Not recorded in southern Africa. *Afrikaans.* Witborsboslyster.

609 Thrush Nightingale *Luscinia luscinia* L 16 cm
This large, drab warbler is usually located by its song as it is very rarely seen, preferring to skulk in tangled undergrowth. If glimpsed, the russet-coloured tail and slightly mottled breast aid identification. *Habitat.* Thick, tangled undergrowth along rivers and in damp areas. Rare, and often overlooked, in the north. *Call.* A rich, warbling song interspersed with harsh, grating notes. *Afrikaans.* Lysternagtegaal.

603 Collared Palm Thrush *Cichladusa arquata* L 19 cm
No other small passerine in the region has the combination of a buff throat and breast narrowly bordered with black, and a bright, whitish eye. The black line encircling the breast is often incomplete, and the sides of the neck and the flanks are washed with grey. Imm. is a mottled buff-brown. *Habitat.* Palm savanna, both the stunted *Hyphaene* and taller *Borassus* palms. Uncommon and highly localized in the extreme north and north-east. *Call.* An explosive whistled song, consisting of 'weet-chuk' or 'cur-lee chuk-chuk' phrases. *Afrikaans.* Palmmôrelyster.

604 Rufoustailed Palm Thrush *Cichladusa ruficauda* L 18 cm
Very similar to Collared Palm Thrush but differs by lacking both the black border on the breast and the grey nape and flanks, and by having a richer rufous crown and back. It also has a darker eye. *Habitat.* Riverine thickets and *Borassus* palm savanna. Status uncertain but extremely rare on the north-western border. *Call.* Not recorded in southern Africa. *Afrikaans.* Rooistertmôrelyster.

607 ▲   609 ▼   605 ▼   608 ▲

603 ▼   604 ▼

620 Whitethroat *Sylvia communis* L 14 cm
Male is readily distinguished by its grey head which contrasts with the silvery white throat and rust-coloured panel in the wings. Female has a brown, not grey head but still shows an obvious white throat and rust-coloured wing panel. *Habitat.* Dry thornveld thickets, often near water. Uncommon summer visitor from Europe to the north of the region. *Call.* A soft 'whit' alarm note and a song which is a harsh, snappy mixture of grating and melodious notes. *Afrikaans.* Witkeelsanger.

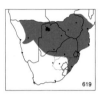

619 Garden Warbler *Sylvia borin* L 14 cm
This small, greyish brown warbler is distinctive only in that it lacks any clear field marks on its head, wings and tail. Its general drabness and uniformity as well as its small, rounded head and indistinct pale eye-ring, aid identification. *Habitat.* Thick tangles in evergreen forests, coastal bush and riverine forests. A common, but easily overlooked, summer visitor from Europe to the central, northern and eastern regions. *Call.* Most often located by its song which is a monotonous warbling interspersed with soft grating phrases. *Afrikaans.* Tuinsanger.

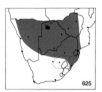

625 Icterine Warbler *Hippolais icterina* L 13 cm
Larger than the similarly coloured Willow Warbler, this species is also normally much brighter yellow below, has a larger bill and a noticeably larger head with a heavily sloped forehead. Distinguished from the Yellow Warbler by having a yellow eye-stripe. *Habitat.* Thornveld, dry broad-leafed woodland and riverine thickets. Common but thinly distributed summer visitor from Europe to the central and northern regions. *Call.* A jumbled song comprising harsh and melodious notes. *Afrikaans.* Spotvoël.

626 Olivetree Warbler *Hippolais olivetorum* L 16 cm
This large grey warbler might be confused with the Fantailed Flycatcher (p. 260) but it lacks the conspicuous white outer tail feathers and the tail-spreading action of that species. The bill is long and heavy, and a pale grey panel is noticeable in the wings. *Habitat.* Dense clumps of thicket in thornveld. Uncommon summer visitor from Europe to Zimbabwe, the Transvaal and Natal. *Call.* Most easily located by its raucous, chattering song, sounding much like a Great Reed Warbler singing from a bush instead of a reedbed. *Afrikaans.* Olyfboomsanger.

643 Willow Warbler *Phylloscopus trochilus* L 11 cm
Distinguished from the Icterine Warbler by its smaller size, much lighter yellow underparts and its thinner, weaker bill. The head shape is sleeker, lacking the steep sloping forehead of the Icterine Warbler. *Habitat.* Frequents a wide range of broad-leafed woodland and dry thornveld habitats. Abundant summer visitor from Eurasia to all parts of the region except the dry west. *Call.* A soft 'hoeet hoeet' and a short melodious song, descending in scale. *Afrikaans.* Hofsanger.

644 Yellowthroated Warbler *Seicercus ruficapillus* L 11 cm
The rufous crown, yellow throat, breast and undertail coverts are diagnostic of this small forest warbler. The belly is grey or off-white and contrasts sharply with the yellow breast and undertail. *Habitat.* Evergreen forests and wooded gullies. Common but localized where suitable forests remain in the south, east and north-east. *Call.* A 'seee suuu seee suuu' song, with variations in phrasing. *Afrikaans.* Geelkeelsanger.

238

619▲

620▲

626▼ 625▼

643▼ 644▼

557 Cape Penduline Tit *Anthoscopus minutus* L 8 cm
Distinguished from eremomelas by its tiny size, rotund body and very short tail. Likely to be confused only with the Grey Penduline Tit from which it differs by having a black forehead, yellowish, not buff belly and flanks, and a speckled throat. *Habitat.* Fynbos, Karoo scrub, semi-desert and dry thornveld. A common endemic to the dry west, with the population extending into southern Angola. *Call.* A soft 'tseep'. *Afrikaans.* Kaapse Kapokvoël.

558 Grey Penduline Tit *Anthoscopus caroli* L 8 cm
Distinguished from Cape Penduline Tit by lacking the black forehead and speckled throat, and by having buffy, not yellowish flanks and belly. Generally it is a much greyer bird than the Cape Penduline Tit. Occurs in small groups of three to five. *Habitat.* Thornveld, and dry broad-leafed and miombo woodlands. Common in the northern, north-eastern and eastern parts of the region. *Call.* A soft 'chissick' or 'tseeep'. *Afrikaans.* Gryskapokvoël.

653 Yellowbellied Eremomela *Eremomela icteropygialis* L 10 cm
The smallest of the eremomelas, it differs from the others by its grey throat and breast, and yellow-washed flanks and belly. Distinguished from similarly coloured white-eyes by its much smaller size, browner upperparts and the lack of white eye-ring. *Habitat.* Thornveld, open broad-leafed woodland and scrub. Common throughout the region except in the south, south-east and north-east. *Call.* Song is a high-pitched, frequently repeated 'tchee-tchee-tchuu'. *Afrikaans.* Geelpensbossanger.

654 Karoo Eremomela (Green Eremomela) *Eremomela gregalis* L 12 cm
The unmarked silvery white underparts distinguish this from other eremomelas. Differs from the Bleating Warbler by having yellow flanks and undertail, greener upperparts and pale eyes. *Habitat.* Karoo and semi-desert scrub, especially along dry rivercourses. An uncommon and localized endemic to the west and south-west. *Call.* A wailing 'quee, quee-quee' song has been described. *Afrikaans.* Groenbossanger.

656 Burntnecked Eremomela *Eremomela usticollis* L 12 cm
When present, the small chestnut throat bar is diagnostic. If the bar is absent, this species can be distinguished from the Greencapped Eremomela by its paler yellow underparts and brown, not green upperparts. *Habitat.* Primarily a thornveld species but also found in mixed, dry broad-leafed woodland and dry rivercourses. Common in suitable areas in the central and northern districts. *Call.* A high-pitched 'chii-cheee-cheee', followed by a sibilant 'trrrrrrrrr'. *Afrikaans.* Bruinkeelbossanger.

655 Greencapped Eremomela *Eremomela scotops* L 12 cm
The bright yellow underparts, greenish upperparts and greenish yellow crown are diagnostic and distinguish this species from other eremomelas and similarly coloured warblers. At close range the pale yellow eye is discernible. *Habitat.* Open broad-leafed woodland and riverine forest. Uncommon and localized in the north and north-east. *Call.* A repeated 'tweer-tweer-tweer'. Small groups utter a harsh 'churr-churr'. *Afrikaans.* Donkerwangbossanger.

240

656▲   558▼   557▼   655▲

653▼   654▼

637 Yellow Warbler *Chloropeta natalensis* L 14 cm
Superficially resembles Icterine and Willow warblers but is much larger than the latter, and is distinguished from both by being a much brighter yellow below and a darker olive green above. This species also shows only a faint trace of an eyebrow stripe and has dark grey legs. Often behaves like a flycatcher, sallying forth from an exposed perch in pursuit of winged insects. *Habitat.* Bracken and briar slopes in mountainous terrain, riverine thickets and forest edges. Locally common in the east. Occurs in coastal scrub in winter. *Call.* A soft 'chip-chip-cheezee-cheeze'. *Afrikaans.* Geelsanger.

627 River Warbler *Locustella fluviatilis* L 13 cm
Easily distinguished from other 'marsh' warblers by its browner upperparts and especially by its graduated, rounded tail. Underparts are buff and there is slight streaking on the breast and the undertail coverts. Differs from Barratt's and Knysna warblers by being much smaller, and by its shorter, less rounded tail. *Habitat.* Thick, tangled growth in riverine scrub, and dam areas. Rare summer visitor to Zimbabwe and the northern Transvaal. *Call.* Easily located by its raspy, insect-like 'derr-derr-zerr-zerr' call. *Afrikaans.* Sprinkaansanger.

642 Broadtailed Warbler *Schoenicola brevirostris* L 17 cm
Easily differentiated from all other warblers by its very long, broad, fan-shaped tail which is black and tipped with buff on the underside. After heavy rains, it is often seen perched on grass stems drying out its tail and wings. *Habitat.* Long rank grass adjoining rivers, dams, and damp areas. Locally common in the east, but easily overlooked. *Call.* A wheezy 'tzzzt-tzzt' and a clear, high-pitched 'peee, peee'. *Afrikaans.* Breëstertsanger.

638 African Sedge Warbler *Bradypterus baboecala* L 17 cm
Distinguished from all other reed-dwelling warblers by its very dark brown coloration, dappled throat and breast, and long, rounded tail. Tends to skulk but does a peculiar display flight over reedbeds. *Habitat.* Reedbeds adjoining dams, lagoons and large rivers. Common in the northern, eastern and southern parts of the region, being absent from the dry central and western regions. *Call.* A low, stuttered 'brrrup brrrup trrp trrp trrp' song and a soft nasal 'nneeeuu'. *Afrikaans.* Kaapse Vleisanger.

639 Barratt's Warbler *Bradypterus barratti* L 15 cm
Very similar to the African Sedge Warbler in appearance but never frequents reedbeds. Its song, more heavily spotted breast, larger size and longer tail distinguish this species from the Knysna Warbler. *Habitat.* Thick tangled growth on the edges of evergreen forests. Locally common in suitable habitat in the eastern and south-eastern mountainous regions. *Call.* A harsh 'chrrrr' alarm note and a 'seee-pllip-pllip' song. *Afrikaans.* Ruigtesanger.

640 Knysna Warbler *Bradypterus sylvaticus* L 14 cm
Slightly smaller than Barratt's Warbler, this species is best distinguished by its shorter, squarer and less heavy tail, reduced spotting on the breast, and by its song. *Habitat.* Well-wooded gullies, and bracken and briar thickets. An uncommon and localized endemic to the south and south-east. *Call.* A loud, clear song which begins with a 'tseep tseep tseep' and increases to a 'trrrrrr' trill towards the end. *Afrikaans.* Knysnaruigtesanger.

641 Victorin's Warbler *Bradypterus victorini* L 16 cm
The most colourful of the *Bradypterus* warblers, it is distinguished by its reddish brown cap contrasting with its yellow eyes, paler throat and buff-red underparts. *Habitat.* Short montane fynbos, especially alongside streams and damp areas. A common but localized endemic to the extreme south. *Call.* Song is diagnostic: a clear, repeated 'weet-weet-weeeo', accelerating towards the end. *Afrikaans.* Rooiborsruigtesanger.

242

640 ▲

627 ▲

641 ▼

642 ▼

637 ▲

639 ▼

638 ▼

628 Great Reed Warbler *Acrocephalus arundinaceus* L 19 cm
Normally located by its harsh, croaking call. Resembles a large African Marsh Warbler, with a long, heavy bill and pale eyebrow stripe. Larger, and darker than Basra Reed Warbler. *Habitat.* Reedbeds surrounding wetlands. Common summer visitor, absent from the south and dry west. *Call.* A prolonged 'chee-chee-chaak-chaak'. *Afrikaans.* Grootrietsanger.

629 Basra Reed Warbler *Acrocephalus griseldis* L 15 cm ☐
Differs from Great Reed Warbler only by smaller size, whiter throat and breast. *Habitat.* Reedbeds and riverine thickets. Rare summer visitor to Moçambique. *Call.* Not recorded in our region. *Afrikaans.* Basrarietsanger.

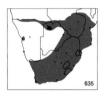

635 Cape Reed Warbler *Acrocephalus gracilirostris* L 17 cm
Much larger than African and European Marsh warblers and has a clearer, whiter throat and breast, and a distinct white eyebrow stripe. The bill is long and heavy and the legs are dark brown. Differs from Greater Swamp Warbler by being clearer white below and by its song. *Habitat.* Reedbeds adjoining wetlands. Common throughout except in the dry central and western sectors. *Call.* A rich, fluty 'cheerup-chee-trrrree' song. *Afrikaans.* Kaapse Rietsanger.

636 Greater Swamp Warbler *Acrocephalus rufescens* L 18 cm
Distinguished from Great Reed Warbler by its smaller size, greyish brown sides to breast, belly and flanks, and darker brown upperparts. Has no white eyebrow stripe. Differs from Cape Reed Warbler by its larger size, darker appearance and its song. *Habitat.* Papyrus swamps. Locally common in the Okavango region. *Call.* A loud 'churrup, churr-churr' interspersed with harsher notes. *Afrikaans.* Rooibruinrietsanger.

634 European Sedge Warbler *Acrocephalus schoenobaenus* L 13 cm
Resembles a cisticola but differs by its short tail which lacks black and white tips. Differentiated from other Acrocephaline warblers by its broad, buffy eyebrow stripe, and streaked crown and back. *Habitat.* Reedbeds bordering wetlands, and thickets, sometimes far from water. Common summer visitor to the north and coastally to the south-east. *Call.* A harsh churring and chattering interspersed with sharp, melodious phrases. *Afrikaans.* Europese Vleisanger.

631 African Marsh Warbler *Acrocephalus baeticatus* L 13 cm
In the field indistinguishable from European Marsh Warbler unless the song is heard. Winter birds are this species as European Marsh Warbler is absent. *Habitat.* Thickets and bush adjoining wetlands. Common summer visitor, with individuals overwintering. Found throughout. *Call.* A harsher, more churring song than that of European Marsh Warbler. *Afrikaans.* Kleinrietsanger.

632 Cinnamon Reed Warbler *Acrocephalus cinnamomeus* L 13 cm ☐
Little-known, and considered a race of African Marsh Warbler. *Habitat.* Same as African Marsh Warbler. Rare in Moçambique and a vagrant to Natal. *Call.* Not recorded in our region. *Afrikaans.* Kaneelrietsanger.

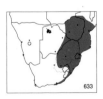

633 European Marsh Warbler *Acrocephalus palustris* L 13 cm
In the field, indistinguishable from African Marsh Warbler unless song is heard. This species usually occurs in drier areas, often far from water. *Habitat.* Bracken and briar on mountain slopes, and forest edges and riverine thickets. Common summer visitor to the east. *Call.* Distinguished from the African Marsh Warbler's song by clear, melodious phrases. Often mimics other birds. *Afrikaans.* Europese Rietsanger.

630 European Reed Warbler *Acrocephalus scirpaceus* L 13 cm ☐
In the field distinguishable from European Marsh Warbler only by song. Doubtfully recorded within our region. *Afrikaans.* Hermanse Rietsanger.

244

628 ▲

633 ▲

631 ▲

636 ▲

635 ▼

634 ▲

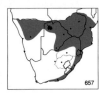

657 Bleating Warbler *Camaroptera brachyura* L 12 cm
Two races occur: a green-backed and a grey-backed race. The grey-backed race might be confused with the Karoo Eremomela but it lacks the yellow flanks and undertail, and has a dark, not pale eye. Both races have a habit of cocking the tail above the back. *Habitat.* Moist, evergreen forests (green-backed), and dry thornveld and broad-leafed woodland (grey-backed). Common in the north, north-east and south-east. *Call.* A nasal 'neeehhh' and a loud, snapping 'bidup-bidup-bidup'. *Afrikaans.* Kwê-kwêvoël.

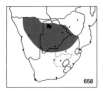

658 Barred Warbler *Camaroptera fasciolata* L 14 cm
Differs from the smaller Stierling's Barred Warbler by having the flanks and undertail coverts washed with warm buff, being less boldly barred below, having a brown, not orange eye, and brown, not flesh-coloured legs. *Habitat.* Dry thornveld and broad-leafed woodland. Common in the dry central and north-western regions. *Call.* A thin 'trrrreee' and 'pleelip-pleelip'. *Afrikaans.* Gebande Sanger.

659 Stierling's Barred Warbler *Camaroptera stierlingi* L 13 cm
Easily distinguished from the larger Barred Warbler by its clear, silvery white underparts barred with black, its orange, not brown eyes and its flesh-coloured, not brown legs. Shows a distinct greenish wash to its upperparts. *Habitat.* Mixed thornveld, open broad-leafed woodland, and miombo and mopane woodland. Common in the eastern and north-eastern regions. *Call.* A fast, breathy 'plip-lip-lip' whistle. *Afrikaans.* Stierlingse Sanger.

660 Cinnamonbreasted Warbler *Euryptila subcinnamomea* L 14 cm
This prinia-like warbler is readily identified by its rust-coloured forehead, breast band, flanks and rump. It cocks its long black tail when alarmed. An agile species, it hops energetically from rock to rock. *Habitat.* Scrub-covered rocky hillsides, in dry river gullies and gorges. An uncommon, localized and very easily overlooked endemic to the south-west. *Call.* A shrill, whistled 'peeeee' or 'chreeee'. Song is a short burst of melodious phrases. *Afrikaans.* Kaneelborssanger.

661 Grassbird *Sphenoeacus afer* L 19 cm
The long, pointed, straggly tail, combined with the chestnut cap and the black malar stripes, is diagnostic. The heavily streaked back and the pointed tail eliminate confusion with the Moustached Warbler. It is much larger than any cisticola in the region. *Habitat.* Coastal and montane fynbos, long rank grass on mountain slopes, and alongside streams. A common endemic to the south, east, and the Eastern Highlands of Zimbabwe. *Call.* A nasal 'pheeeoo' and a jangled, musical song. *Afrikaans.* Grasvoël.

662 Rockrunner (Damara Rockjumper) *Achaetops pycnopygius* L 17 cm
Although this species is also heavily streaked above, it is distinguished from the Grassbird by its shorter, rounded tail, by lacking the chestnut cap, and by having a chestnut belly and flanks. Scrambles swiftly over rocks to dive for cover in long grass. *Habitat.* Boulder-strewn grassy hillsides and the bases of small hills. A common but easily overlooked endemic to Namibia. *Call.* A hollow, melodious warbling 'rooodle-trrooodlee'. *Afrikaans.* Rotsvoël.

663 Moustached Warbler *Melocichla mentalis* L 19 cm
This large warbler might be mistaken for the Grassbird but it is easily differentiated by its shorter, rounded tail and uniform, not streaked back. Distinguished from the smaller Broadtailed Warbler by having black malar stripes and by lacking any buff tips to its undertail feathers. *Habitat.* Long rank grass adjoining forests and in open glades, often near streams or damp areas. Rare and localized in Zimbabwe and Moçambique. *Call.* Song described as 'tip-tiptwiddle-iddle-see'. *Afrikaans.* Breëstertgrasvoël.

658 ▲

659 ▲

660 ▲

662 ▼

657 ▲ 661 ▲

663 ▼

621 Titbabbler *Parisoma subcaeruleum* L 15 cm
Paler grey than the very similar Layard's Titbabbler, this species is best distinguished by its bright chestnut vent, by having the white on the tail confined to the tip and by dark outer tail feathers. The streaking on the throat is bolder and more extensive than in Layard's Titbabbler. *Habitat.* Dry thornveld, and thickets and dry scrub in semi-desert. Common throughout except for the moister eastern regions. *Call.* A loud, fluty 'cheruuup-chee-chee', interspersed with harsh chatters. *Afrikaans.* Bosveldtjeriktik.

622 Layard's Titbabbler *Parisoma layardi* L 15 cm
Generally much darker than the Titbabbler. Best distinguishing character is the white, not chestnut vent. The silvery white eye contrasts with the dark head, and the throat streaking is less pronounced than in the Titbabbler. Superficially resembles the Fantailed Flycatcher but lacks the tail-spreading action of that species and has streaking on the breast. *Habitat.* Dry thornveld regions, where it is frequently found together with the Titbabbler. A common but localized endemic to the west and south. *Call.* A clear 'pee-pee-cheeri-cheeri', similar in quality to that of the Titbabbler but with different phrasing. *Afrikaans.* Grystjeriktik.

706 Fairy Flycatcher *Stenostira scita* L 12 cm
This small flycatcher, which behaves more like a warbler, shows a pinkish grey wash across the breast, a black stripe through the eye topped by a broad white eyebrow stripe, a white stripe in the wing and white outer tail feathers. Quite unlike any other warbler or flycatcher in the region. *Habitat.* Dry Karoo bush and scrub, montane fynbos, and scrub-filled mountain gullies. Uncommon endemic to the central, southern and western regions.
Call. A repeated, wispy 'tisee-tchee-tchee' phrase and a descending 'cher cher cher'. *Afrikaans.* Feevlieëvanger.

651 Longbilled Crombec *Sylvietta rufescens* L 12 cm
This small, plump warbler which appears to be almost tailless, is distinguished by its slightly decurved bill. It lacks the chestnut breast band and ear patches· of the Redcapped Crombec and differs from the Redfaced Crombec by having a longer bill, brownish, not ashy grey upperparts and by lacking the russet tinge to the face. *Habitat.* Frequents a wide range of woodlands, preferably dry, and is also found in desert scrub. Avoids dense evergreen and coastal forests. Common and found throughout most of the region.
Call. A repeated 'trree-trrriit, trree-trrriit' and a harsher 'ptttt'. *Afrikaans.* Bosveldstompstert.

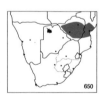

650 Redfaced Crombec *Sylvietta whytii* L 11 cm
Distinguished from the Longbilled Crombec by having pale ashy grey upperparts, a noticeably shorter bill, a pale chestnut face and richer, buffier underparts. Differs from the Redcapped Crombec by lacking that species' distinctive chestnut ear patches and breast band. *Habitat.* Miombo woodland, often joining mixed bird parties. Locally common in Zimbabwe and Moçambique. *Call.* A thin 'si-si-si-see' has been recorded.
Afrikaans. Rooiwangstompstert.

652 Redcapped Crombec *Sylvietta ruficapilla* L 12 cm
Easily distinguished from the other two crombec species by its diagnostic rufous ear patches and breast band. The forehead and crown are a darker chestnut and the back is a paler ashy grey than that of the Redfaced Crombec. *Habitat.* Miombo woodland. Rare in our region, with scattered sight records from the Zambezi River in Zimbabwe. *Call.* Not recorded in our area. *Afrikaans.* Rooikroonstompstert.

706 ▲   622 ▼   652 ▼   621 ▲

651 ▼   650 ▼

645 **Barthroated Apalis** *Apalis thoracica* L 13 cm
The only apalis in the region to show both white outer tail feathers and a pale eye. Racially very variable in colour and best distinguished from the similar Rudd's Apalis by the white outer tail feathers and the pale eye, and by lacking the small white eyebrow stripe. *Habitat.* Evergreen and coastal forests, scrub and fynbos. Common in the central, eastern and southern regions. *Call.* A sharp, rapidly repeated 'pilllip-pilllip-pilllip'. *Afrikaans.* Bandkeelkleinjantjie.

647 **Blackheaded Apalis** *Apalis melanocephala* L 14 cm
The only apalis in the region to have black upperparts and white underparts. At close range the white eye, which contrasts strongly with the black cap, is visible. *Habitat.* Usually found high in the canopy of evergreen, coastal and tall riverine forests. Common but localized in Moçambique and the highlands of Zimbabwe. *Call.* A piercing 'wiii-tiiit-wiii-tiiit', repeated many times. *Afrikaans.* Swartkopkleinjantjie.

648 **Yellowbreasted Apalis** *Apalis flavida* L 13 cm
The combination of grey head, white throat, yellow breast (sometimes with a small black lower band), and white belly, is diagnostic. The back is a much paler leaf green colour than that of other apalises. This species is more slender and longer tailed than any eremomela. *Habitat.* Frequents a wide range of woodland habitats but avoids montane evergreen forests. Common in the extreme north, east and south-east. *Call.* A fast, buzzy 'chizzick-chizzick-chizzick'. *Afrikaans.* Geelborskleinjantjie.

649 **Rudd's Apalis** *Apalis ruddi* L 13 cm
Could be confused with the Barthroated Apalis but lacks the white outer tail feathers, has a small white stripe in front of and above the eye, and a dark, not pale eye. The lime-green back contrasts strongly with the grey head and the bird looks altogether much 'neater' than the Barthroated. *Habitat.* Thornveld and coastal forest. A common but localized endemic to Natal and Moçambique. There are recent records from Malawi. *Call.* When one bird calls its fast 'tuttuttuttut', another responds with a slower 'clink-clink-clink'. *Afrikaans.* Ruddse Kleinjantjie.

646 **Chirinda Apalis** *Apalis chirindensis* L 14 cm
Could be confused with the Whitetailed Flycatcher which occurs in the same area but has a slender, not broad, fan-shaped tail, and completely different habits. A restless bird, it creeps and flits through the canopy when foraging and does not fan its tail or swing from side to side on a perch. No other apalis in the region is uniformly grey in colour. *Habitat.* Evergreen forests and their edges. An uncommon and localized endemic to the highland forests of Zimbabwe and Moçambique. *Call.* A sharp 'chipip chipip' has been recorded. *Afrikaans.* Gryskleinjantjie.

647▲   646▲   648▼

645▼ 649▼

664 Fantailed Cisticola *Cisticola juncidis* L 10 cm
One of the most abundant and commonly seen of the very small cisticolas. In its display, the male flies at a height of 5 to 20 m in a bounding, erratic manner. It does not 'snap' its wings. The tail is more boldly marked than in other small cisticolas: black above and below, and broadly tipped with white. *Habitat.* Grasslands, fallow lands and damp, rank grassy areas. Common and widely distributed, except in the dry west. *Call.* A repeated 'zit, zit, zit' or a faster 'chit-chit-chit' is given in flight. *Afrikaans.* Landeryklopkloppie.

665 Desert Cisticola *Cisticola aridula* L 10 cm
Similar to Fantailed Cisticola but has a very different display and its tail is less boldly marked. Often sings when perched on a grass stem but mostly during its low-level display flight which lacks the bounding, erratic movements shown by the Fantailed. *Habitat.* Grasslands in dry regions. Common and widely distributed but absent from the south-west, east and north-east. *Call.* Song is a 'zink zink zink' or 'sii sii sii sii'. When alarmed it utters a 'tuc tuc tuc tuc', and snaps its wings. *Afrikaans.* Woestynklopkloppie.

666 Cloud Cisticola *Cisticola textrix* L 10 cm
Birds in the southern Cape are easily recognized because of streaking on the breast. Elsewhere, almost indistinguishable from Ayres' and Desert cisticolas except that this species has a stocky, plump body and unusually long legs and feet. *Habitat.* Upland and coastal grasslands. Common and widespread in the southern, south-eastern and central regions. *Call.* Song, a 'see-see-see-see-chick-chick-chick', is delivered at great height. Does not snap its wings before landing. *Afrikaans.* Gevlekte Klopkloppie.

667 Ayres' Cisticola *Cisticola ayresii* L 10 cm
Almost indistinguishable from Cloud Cisticola except when heard and seen in display flight. Appears neater and slimmer than Cloud Cisticola, and its legs and feet do not appear unusually long. *Habitat.* Upland grasslands and occasionally near the coast. Common and widely distributed in the uplands of the south-eastern and eastern central regions. *Call.* Song, a 'soo-see-see-see', is given at a great height. On descending and just before it jinks, loudly snaps its wings many times. *Afrikaans.* Kleinste Klopkloppie.

668 Palecrowned Cisticola *Cisticola brunnescens* L 10 cm
Male in breeding plumage is easily distinguished from other very small cisticolas as it has a black bill and black lores contrasting with a pale, buffy crown. In non-breeding plumage it is indistinguishable from Cloud and Ayres' cisticolas. *Habitat.* Frequents damp or marshy areas in upland grasslands. Uncommon and highly localized in the east. *Call.* Display flight is undertaken at both high and low levels. Song is a soft, hardly discernible 'tsee-tsee-tsee-tititititititi'. Does not wing snap. *Afrikaans.* Bleekkopklopkloppie.

681 Neddicky *Cisticola fulvicapilla* L 11 cm
The greyish underparts, uniform brownish upperparts and bright chestnut cap render the southern race one of the easiest cisticolas to identify. The northern race is similar to Shortwinged Cisticola but has buffier underparts, a chestnut cap and a longer tail. *Habitat.* Diverse. Common and widely distributed, but absent from the Karoo and dry west. *Call.* Song is a soft, breathy 'cheerup-cheerup-cheerup'. Alarm call is a fast 'tictictictic'. *Afrikaans.* Neddikkie.

680 Shortwinged Cisticola *Cisticola brachyptera* L 10 cm
Resembles Neddicky but is shorter tailed and has clear buffy white underparts. Upperparts are a uniform dark brown. *Habitat.* Grassland clearings in miombo woodland. Uncommon and localized in Zimbabwe and Moçambique. *Call.* Display flight is accompanied by a soft 'see-see-sippi-sippi'. *Afrikaans.* Kortvlerktinktinkie.

680 ▲

666 ▼   664 ▲

665 ▲   667 ▼

668 ▼

681 ▼

669 Greybacked Cisticola *Cisticola subruficapilla* L 13 cm
The southern race has a diagnostic grey back finely streaked with black, but the northern race is very difficult to distinguish from the Wailing Cisticola. The main difference lies in the underpart coloration which, in this species, is a cold greyish buff, not the warm buff on the belly and flanks of the Wailing Cisticola. *Habitat.* Fynbos, grassy hillsides and desert scrub. A common endemic to the drier southern and western regions. *Call.* A muffled 'prrrrrrt' and sharp 'phwee phweee' notes. *Afrikaans.* Grysrugtinktinkie.

670 Wailing Cisticola *Cisticola lais* L 13 cm
Distinguished from the northern race of Greybacked Cisticola by having much warmer buff, not cold grey-buff belly and flanks. The calls and songs of these two species are very similar. *Habitat.* Grassy, bracken and briar mountain slopes and gullies. Common in the southern, eastern and central eastern regions. *Call.* Very mixed and variable and not readily distinguishable from that of the Greybacked Cisticola. *Afrikaans.* Huiltinktinkie.

679 Lazy Cisticola *Cisticola aberrans* L 14 cm
In shape and habits, the most prinia-like cisticola of the region but it differs from any prinia by its rust-coloured crown and warm buffy underparts. Distinguished from the Neddicky by its long thin tail which is often held cocked. *Habitat.* Grass- and bush-covered hillsides strewn with rocks and boulders. Common but localized in the south-east and central regions, and in Zimbabwe. *Call.* Song is a 'tzeeee-tzeeeh-cheee-cheee'. *Afrikaans.* Luitinktinkie.

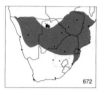

672 Rattling Cisticola *Cisticola chiniana* L 13 cm
The most obvious and abundant cisticola of the thornveld. Easily confused in plumage coloration with the Tinkling Cisticola but has much less red on the head, and the tail is not as brightly rust-coloured. *Habitat.* Thornveld, dry broad-leafed woodland and edges of coastal forest. Common and widely distributed in the north and east. *Call.* Song is a loud 'chueee-chueee-cherrrrrr', with variations on this phrase. *Afrikaans.* Bosveldtinktinkie.

671 Tinkling Cisticola *Cisticola rufilata* L 14 cm
Where their distribution ranges overlap, this species is most easily confused with the Rattling Cisticola. Differs by being much redder on the head and tail. It also shows a clear white eyebrow stripe. Song is diagnostic. *Habitat.* Dry thornveld and mopane woodland. Common in the central and north-western parts of the region. *Call.* A series of soft, tinkling, bell-like notes. Alarm call is rendered as 'dee-dee-du-du'. *Afrikaans.* Rooitinktinkie.

678 Croaking Cisticola *Cisticola natalensis* L 13-17 cm
This large cisticola is unlikely to be confused with any other because of its size, bulky body shape and short tail, and its unusually thick bill. Upperparts are well mottled with dark brown and there is no rufous on the head. The female is considerably smaller than the male. *Habitat.* Grassland in bush clearings, and grassy hillsides with scattered bush. Common but localized in the north-east, east and south-east. *Call.* A deep 'trrrrp' or 'chee-fro' is given during its bounding display flight or from an exposed perch. *Afrikaans.* Groottinktinkie.

671▲

679▲

678▲

670▼

672▲

669▼

677 Levaillant's Cisticola *Cisticola tinniens* L 14 cm
Could be confused with the Blackbacked Cisticola but can normally be distinguished by its reddish, not grey tail and, at close range, it can be seen that the black feathers of its back are edged with brown, not grey. Its song and call are very different to those of the Blackbacked Cisticola. *Habitat.* Reedbeds and long grass adjacent to rivers and dams. Common in the eastern half of the region but absent from the north-west. *Call.* A musical 'chrip-trrrup-trreee' and a wailing 'cheee-weee-weee'. *Afrikaans.* Vleitinktinkie.

675 Blackbacked Cisticola *Cisticola galactotes* L 13 cm
Most easily confused with Levaillant's Cisticola which also has a black-streaked back. However, this species can be readily distinguished by its call and as its back is more boldly streaked with black. When breeding it shows a greyish, not red tail. *Habitat.* Reedbeds, long grass and sedges near water. Common in the eastern and north-central regions. *Call.* A loud, harsh 'tzzzzzzrrp' and a louder, whistled alarm 'prrrt'. *Afrikaans.* Swartrugtinktinkie.

676 Chirping Cisticola *Cisticola pipiens* L 15 cm
This species could be confused with the similar Blackbacked Cisticola: their distribution ranges overlap and they are often found in association. However, it differs in call, by having its back less boldly streaked with black, and by having a buffish wash on its breast and belly. *Habitat.* Reedbeds and papyrus swamps. Common but very localized in suitable habitat in the Okavango Delta. *Call.* A loud 'cheet-cheet-zrrrrr' and 'chwer-chwer-chwer'. *Afrikaans.* Piepende Tinktinkie.

674 Redfaced Cisticola *Cisticola erythrops* L 13 cm
Differs from other reed-dwelling cisticolas by having a uniformly coloured, not black-streaked back. The face, sides of breast and flanks are more obviously washed with russet in non-breeding plumage. Differs from the Singing Cisticola by lacking russet edges to its primaries. *Habitat.* Tall stands of reedbeds, bulrushes and papyrus. Common in the north-eastern and eastern sections of the region. *Call.* A series of piercing whistles 'weee, cheee, cheee, cheer, cheer', rising and falling in scale. *Afrikaans.* Rooiwangtinktinkie.

673 Singing Cisticola *Cisticola cantans* L 13 cm
Distinguished from the similar Redfaced Cisticola by having diagnostic rufous edges to the primaries and a rufous cap contrasting with the uniformly grey-brown back. The red on the crown does not envelop the face and ear coverts as it does in the Redfaced Cisticola. *Habitat.* Long grass in clearings in miombo woodlands. Uncommon and highly localized in Zimbabwe and Moçambique. *Call.* Given from an exposed perch and rendered as 'jhu-jee' or 'wheecho'. *Afrikaans.* Singende Tinktinkie.

688 Rufouseared Warbler *Malcorus pectoralis* L 15 cm
The reddish ear coverts and narrow black breast band are diagnostic. Very prinia-like in appearance and habits, although it forages freely on the ground and often runs swiftly from bush to bush. *Habitat.* Fynbos, and Karoo and semi-desert scrub. A common and widely distributed endemic to the south and west. *Call.* A harsh, scolding 'chweeo, chweeo, chweeo'. *Afrikaans.* Rooioorlangstertjie.

675▲ 673▼ 688▼ 674▲

676▼ 677▼

682 Redwinged Warbler *Heliolais erythroptera* L 13 cm
This long-tailed warbler resembles the Tawnyflanked Prinia but is easily distinguished by its bright rufous wings which contrast with dark brown upperparts. In non-breeding plumage the upperparts assume a rusty colour, blending with the rufous wings. *Habitat.* Long grass in woodland clearings and alongside streams. Uncommon and localized in the north-east.
Call. Described as a 'pseep-pseep-pseep'. *Afrikaans.* Rooivlerksanger.

683 Tawnyflanked Prinia *Prinia subflava* L 11 cm
Distinguished from the non-breeding Blackchested Prinia by having a white, not yellow throat and breast, warm buff flanks and belly, and russet edges to the wings. Non-breeding Roberts' Prinia is similar but is generally much darker and lacks the russet edges to its wings. *Habitat.* Thick, rank vegetation adjoining rivers and dams. Common in the northern and eastern sectors of the region. *Call.* A rapidly repeated 'przzt-przzt-przzt' and a harsh 'chrzzzt'. *Afrikaans.* Bruinsylangstertjie.

684 Roberts' Prinia *Prinia robertsi* L 14 cm
Unmistakable in breeding plumage when its throat and breast are washed with grey and its upperparts appear very dark. In non-breeding plumage it might be confused with the Tawnyflanked Prinia but it is much darker above, has a pale eye and lacks the russet edges to the wings. *Habitat.* Thick bracken and briar adjoining forests. An uncommon and highly localized endemic to the highlands of Zimbabwe and Moçambique. *Call.* A loud, chattering 'cha-cha-cha-cha'. *Afrikaans.* Woudlangstertjie.

685 Blackchested Prinia *Prinia flavicans* L 15 cm
The only prinia in the region to show a broad black chest band. This band is absent in the non-breeding plumage and the bird might then be mistaken for a Tawnyflanked Prinia, but it is very yellow below and sometimes shows the semblance of a spotted breast band. *Habitat.* Generally very arid scrub and thornveld. A common endemic to the central and north-western sectors.
Call. A series of drawn-out 'zzzrt-zzzzrt-zzzrt-zzzrt' notes.
Afrikaans. Swartbandlangstertjie.

686 Spotted Prinia (Karoo Prinia) *Prinia maculosa* L 14 cm
Easily distinguished from other prinias, except the Namaqua, by its extra-long tail and by the conspicuous streaking on its underparts. Where its range overlaps with that of the Namaqua Prinia confusion may arise, but that species lacks buff tips to its undertail and has very faint streaking confined to the breast. *Habitat.* Fynbos, Karoo scrub and bracken-covered slopes in mountainous terrain. Common endemic to the south and east. *Call.* A sharp 'chleet-chleet-chleet' and a faster 'tit-tit-tit-tit'. *Afrikaans.* Karoolangstertjie.

687 Namaqua Prinia *Prinia substriata* L 14 cm
Likely to be confused only with the Spotted Prinia but it differs by having a more russet-coloured back, a longer, wispier tail, and very faint streaking confined to the breast. Unlike the Spotted Prinia, it lacks buff tips to its undertail feathers. *Habitat.* Thick bush in dry river gullies, and reedbeds adjoining rivers and dams. A common but thinly distributed endemic to the south-west. *Call.* A high-pitched 'trreep-trreep-trrrrrrr' song.
Afrikaans. Namakwalangstertjie.

684▼ 682▲

686▲ 683▼ 687▼

685▼

689 Spotted Flycatcher *Muscicapa striata* L 14 cm
Larger than the similar Dusky Flycatcher, this species is also paler below with more definite streaking and blotching on its forehead and underparts. If seen, the streaked forehead is diagnostic. *Habitat.* Frequents a wide range of habitats, from the edges of evergreen forests to semi-arid bush. Common summer visitor from Eurasia to most parts of the region. *Call.* An almost inaudible 'tzee'. *Afrikaans.* Europese Vlieëvanger.

690 Dusky Flycatcher *Muscicapa adusta* L 12 cm
Smaller and much darker than the Spotted Flycatcher. At close range, the lack of forehead streaking, which distinguishes these two species, becomes obvious. The chin is pale and unmarked, the underparts washed grey-brown with ill-defined streaking. *Habitat.* Evergreen forest edges and glades, and the larger riverine forests. Common in the east and south. *Call.* A soft, high-pitched 'tzzeet' and 'tsirit'. *Afrikaans.* Donkervlieëvanger.

692 Collared Flycatcher *Ficedula albicollis* L 12 cm
In its pied breeding plumage, the male of this small flycatcher is unmistakable. It superficially resembles the male Fiscal Flycatcher but that species is much larger and longer tailed, and has no white collar. Female and non-breeding male superficially resemble female Mashona Hyliota but differ by being greyer above and by having a slight grey, not yellow wash below. Displays typical flycatcher habits. *Habitat.* Miombo woodland, and open, mixed thornveld. Extremely rare summer visitor to Zimbabwe and the Transvaal. *Call.* Not recorded in our region but described as 'whit-whit'. *Afrikaans.* Withalsvlieëvanger.

691 Bluegrey Flycatcher *Muscicapa caerulescens* L 15 cm
In comparison with the Fantailed Flycatcher, this species is much bluer grey above, greyer below, and it lacks the white outer tail feathers. Differs too by its typical aerial flycatching habits, unlike the Fantailed Flycatcher which forages while creeping around in the foliage. *Habitat.* Evergreen and riverine forests and moist, open broad-leafed woodland. Common in the east and north-east. *Call.* Song is a soft 'sszzit-sszzit-sreee-sreee'. *Afrikaans.* Blougrysvlieëvanger.

693 Fantailed Flycatcher (Leadcoloured Flycatcher)
Myioparus plumbeus L 14 cm
Distinguished from the similar Bluegrey Flycatcher by its habit of frequently raising and fanning its tail over its back, and in so doing, displaying its conspicuous white outer tail feathers. It is also differentiated by its creeping, not aerial habit of feeding. *Habitat.* Riverine forest, mixed thornveld and open broad-leafed woodland. Uncommon and localized in the north and east. *Call.* A tremulous, whistled 'treee-trooo'. *Afrikaans.* Waaierstertvlieëvanger.

689 ▲

690 ▼

691 ▼

692 ▲

693 ▼

694 Black Flycatcher *Melaenornis pammelaina* L 22 cm
Distinguished from the male Black Cuckooshrike (p. 212) by lacking the yellow base to its bill and by having a square, not rounded tail. Differs from the Squaretailed Drongo (p. 212) by having a square-ended, un-notched tail and, at close range, its brown, not red eye can be seen. Hunts insects in typical flycatcher manner. *Habitat.* Mixed woodland, thornveld and forest edges. Common in the north, east and south-east. *Call.* Song is a 'tzzit-terra-loora-loo'. *Afrikaans.* Swartvlieëvanger.

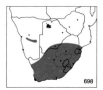

698 Fiscal Flycatcher *Sigelus silens* L 17-20 cm
Often mistaken for the Fiscal Shrike (p. 278) but is easily recognized in flight as it has a shorter tail with conspicuous white patches on the sides. It can also be distinguished by its bill which is thin and flat, not stubby and hooked, and by the white in its wings being confined to the secondaries, not the shoulders. It is almost twice the size of the Collared Flycatcher. *Habitat.* Open broad-leafed woodland and bush, and has also adapted to suburbia. A common endemic to the southern part of the region. *Call.* A weak, soft, chittering song and a 'tssisk' alarm call. *Afrikaans.* Fiskaalvlieëvanger.

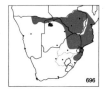

696 Mousecoloured Flycatcher (Pallid Flycatcher) *Melaenornis pallidus* L 17 cm
The eastern counterpart of the Marico Flycatcher, this nondescript bird differs by having buffish brown underparts and by lacking the pale buffish panel in the wings. Its range does not overlap with that of the much larger Chat Flycatcher. *Habitat.* Moist, open broad-leafed woodland and mixed thornveld. Common and widespread in the north and east. *Call.* Song is a melodious warbling interspersed with harsh chitters. Alarm call is a soft 'churr'. *Afrikaans.* Muiskleurvlieëvanger.

695 Marico Flycatcher *Melaenornis mariquensis* L 18 cm
Its clear white underparts and buffish wing panel differentiate this species from the Mousecoloured Flycatcher. Imm. is heavily spotted with buff above and streaked with brown below. Sallies forth from roadside fenceposts and trees to catch insects. *Habitat.* Mixed thornveld and open, dry broad-leafed woodland. Common endemic to the dry north-west. *Call.* Song is a soft 'tsii-cheruk-tukk'. *Afrikaans.* Maricovlieëvanger.

697 Chat Flycatcher *Melaenornis infuscatus* L 20 cm
A large flycatcher of thrush-like proportions. It is uniform brownish above with paler underparts and a pale brown panel on folded secondaries. Imm. is heavily spotted with buff above and below. *Habitat.* Fynbos, and Karoo and desert scrub. A common but thinly distributed endemic to the western sector of the region. *Call.* Song is a rich, warbled 'cher-cher-cherrip', with squeaky, hissing notes. *Afrikaans.* Grootvlieëvanger.

618 Herero Chat *Namibornis herero* L 17 cm
At a glance might be mistaken for a Familiar Chat but has a striking head pattern: the black line which runs through the eye contrasts with the clear white eyebrow stripe. The outer tail feathers and rump are rust-coloured and only at close range can the faint streaking be seen on the breast. *Habitat.* Dry scrub and thornveld at the base of hills and in boulder-strewn country. An uncommon and thinly distributed endemic to the north-west. *Call.* Mostly silent except during the breeding season when a melodious, warbling song is uttered. *Afrikaans.* Hererospekvreter.

696 ▲

695 ▲

698 ▼

694 ▲

697 ▲

618 ▼

700 Cape Batis *Batis capensis* L 13 cm

Male is the only batis in the region with completely russet-washed wings and flanks. Female also has distinctive russet wing coverts and differs from the smaller female Woodward's Batis by lacking the complete white eyebrow stripe. *Habitat.* Moist evergreen forests and heavily wooded gorges in mountain ranges. Common endemic with scattered distribution in the south, east and north-east. *Call.* A soft 'chewrra-warrra-warrra' and 'foo-foo-foo'. *Afrikaans.* Kaapse Bosbontrokkie.

703 Pririt Batis *Batis pririt* L 12 cm

Although the male is very similar to the male Chinspot Batis, their calls differ and their ranges do not overlap. The female differs from the female Chinspot Batis by having a rust-coloured wash over the throat and breast, not the clearly defined chestnut chin patch. *Habitat.* Dry thornveld, broad-leafed woodland and dry riverine bush. Common endemic to the dry western region. *Call.* A series of slow 'teuu, teuu, teuu, teuu' notes, descending in scale. *Afrikaans.* Priritbosbontrokkie.

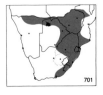

701 Chinspot Batis (Whiteflanked Batis) *Batis molitor* L 13 cm

The female has a distinctive, clearly defined rufous spot on the chin, unlike the suffused chestnut on other female batises' throats. The male has clear, unmarked white flanks which distinguish it from all other male batises except the Pririt: however, their ranges do not overlap. *Habitat.* Prefers drier broad-leafed woodland than the Cape Batis and is found in dry thornveld. Common in the northern and eastern sectors. *Call.* A clear 'teuu-teuu-teuu', and harsh 'chrr-chrr' notes. *Afrikaans.* Witliesbosbontrokkie.

704 Woodward's Batis *Batis fratrum* L 12 cm

Male lacks a black breast band and, although it resembles the female Pririt Batis, their distribution ranges are mutually exclusive. Female is very similar to the female Chinspot Batis but lacks the obvious chestnut chin patch and has rust-coloured wing panels. *Habitat.* Coastal forest and scrub. Locally common in the north-east. *Call.* A clear, penetrating whistle 'tch-tch-pheeeoooo'. *Afrikaans.* Woodwardse Bosbontrokkie.

702 Mozambique Batis *Batis soror* L 10 cm

The smallest batis in the region. The male is distinguished from the male Chinspot Batis by having a narrower black breast band, black-flecked flanks and a dappled black-and-white back. The female is much smaller than the female Chinspot Batis and the chestnut chin patch is ill-defined. *Habitat.* Miombo woodland and coastal forest. Common in northern Moçambique. *Call.* A soft, frequently repeated 'tcheeo, tcheeo, tcheeo'. *Afrikaans.* Mosambiekbosbontrokkie.

700 ▲

703 ▲   **704 ▼**

702 ▼   **701 ▲**

624 Mashona Hyliota *Hyliota australis* L 14 cm
Distinguished from the very similar Yellowbreasted Hyliota by having matt blue-black upperparts, a small white panel on the secondary coverts not extending on to the tertials, and paler yellow underparts. Female has warm brown, not grey-brown upperparts as in the female Yellowbreasted Hyliota. Both *Hyliota* species are distinguished from the Collared Flycatcher by having shorter tails and by their habit of creeping about the canopy in search of food. *Habitat.* Miombo and mopane woodland. Confined to the north-eastern part of the region. *Call.* Trilling and chittering similar to that of the Yellowbreasted Hyliota. *Afrikaans.* Mashonahyliota.

623 Yellowbreasted Hyliota *Hyliota flavigaster* L 14 cm
Male distinguished from male Mashona Hyliota by having distinctly glossy blue-black upperparts, a greater extent of white in the wings and richer yellow underparts. Female difficult to distinguish from female Mashona Hyliota but has grey-brown, not warm brown upperparts. *Habitat.* Miombo and mopane woodland. Uncommon in the north-eastern sector of the region. *Call.* A high-pitched 'trreet trreet'. *Afrikaans.* Geelborshyliota.

707 Livingstone's Flycatcher *Erythrocercus livingstonei* L 12 cm
Unmistakable small flycatcher which has a long rufous tail with a black subterminal band, sulphur yellow underparts and a blue-grey cap. Continually in motion, flicking and fanning its tail sideways. *Habitat.* Riverine and coastal forests. Locally common in Zimbabwe and Moçambique. *Call.* A sharp 'chip-chip' and a clear, warbled song. *Afrikaans.* Rooistertvlieëvanger.

796 Cape White-eye *Zosterops pallidus* L 12 cm
Usually distinguished from the Yellow White-eye by its greyish underparts and green, not yellow upperparts. However, birds in the northern regions have pale green, almost yellow backs; in which case, only the greyish green vent and greenish, not yellow head differentiate this species from the Yellow White-eye. *Habitat.* A ubiquitous endemic, but absent from extreme desert regions and the north, where it is replaced by the Yellow White-eye. *Call.* Small groups call a continual 'tweee-tuuu-twee-twee'. *Afrikaans.* Kaapse Glasogie (Groenglasogie).

797 Yellow White-eye *Zosterops senegalensis* L 11 cm
The bright sulphur yellow underparts and very pale green, almost yellow upperparts distinguish this species from the Cape White-eye. The head of this species is almost completely yellow whereas that of the Cape White-eye is green. The imm. is a very much paler yellow and green than the imm. Cape White-eye. *Habitat.* Found in a variety of woodlands, from thornveld to montane forests. Common in the north-east. *Call.* Very similar in quality to that of the Cape White-eye but this species has a faster delivery. *Afrikaans.* Geelglasogie.

266

707▲   624▼   623▲

797▼   796▼

710 **Paradise Flycatcher** *Terpsiphone viridis* L 23 cm (plus 18 cm tail)
The dark head and breast, bright blue bill and eye-ring, and chestnut back and tail, render this bird easily identifiable. The male in breeding plumage has an extraordinarily long tail and is unmistakable. Generally, this species is very active, dashing to and fro amongst leafy foliage. *Habitat.* Evergreen, coastal and riverine forests, and mixed thornveld. Common except in the dry central and western regions. *Call.* Harsh 'tic-tic-chaa-chaa' notes, similar to those of the Bluemantled Flycatcher. Song is a loud 'twee-tiddly-te-te'. *Afrikaans.* Paradysvlieëvanger.

709 **Whitetailed Flycatcher** *Trochocercus albonotatus* L 15 cm
A nondescript, dark-headed flycatcher easily recognized by its unusual habit of closing and fanning its tail, thereby displaying white outer tail feathers and white spots on the tips of the tail, while moving from side to side on a branch. The female Bluemantled Flycatcher does not exhibit this action and lacks conspicuous white in the tail, therefore confusion between these two species is unlikely to arise. *Habitat.* Well-forested gorges and evergreen montane forests. Locally common and found only in the mountains of Zimbabwe and Moçambique. *Call.* Song is a fast 'tsee-tsee-teuu-choo' and other jumbled notes. *Afrikaans.* Witstertvlieëvanger.

708 **Bluemantled Flycatcher** *Trochocercus cyanomelas* L 18 cm
The glossy black head and shaggy crest render the male unmistakable. The female could possibly be confused with the Whitetailed Flycatcher but is larger, paler grey on the head and throat, has a white wing stripe and lacks white in the tail. *Habitat.* Evergreen, coastal and riverine forests. Common in the south, east and north-east. *Call.* Harsh 'tic-tic-chaaa-chaaa' notes and a fluty, whistled song, similar to that of the Paradise Flycatcher. *Afrikaans.* Bloukuifvlieëvanger.

699 **Vanga Flycatcher (Black-and-White Flycatcher)** *Bias musicus* L 16 cm
Male is unmistakable with its diagnostic black crest, black throat and bib, and small white wing patch. Female has a black cap with a slight crest, and a bright chestnut back and tail. Male is distinguished from the smaller, female Wattle-eyed Flycatcher by lacking the red wattle over the eye. *Habitat.* Mixed miombo woodland, coastal forest and mangrove stands. Uncommon and thinly distributed in northern Moçambique. *Call.* Song is a 'whitu-whitu-whitu' and the alarm note is a sharp 'we-chip'. *Afrikaans.* Witpensvlieëvanger.

705 **Wattle-eyed Flycatcher (Blackthroated Wattle-eye)**
Platysteira peltata L 18 cm
In the male, the conspicuous, bright red wattle over the eye, the black cap and narrow black breast band are diagnostic. The female is distinguished from the male Vanga Flycatcher by showing red eye wattles, a pale rump and by lacking the crest. *Habitat.* Coastal and riverine forests and, occasionally, mangrove stands. Uncommon and localized in the east and north-east. *Call.* A repeated 'wichee-wichee-wichee-wichee'. *Afrikaans.* Beloogbosbontrokkie.

710▲   699▲

708▲   709▼   705▼

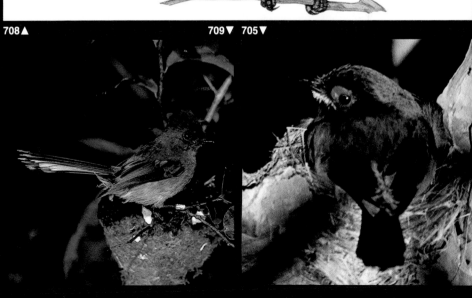

711 **African Pied Wagtail** *Motacilla aguimp* L 20 cm

Unmistakable large wagtail with black and white pied plumage. Imm. might be confused with Cape Wagtail but is distinguished by its white wing coverts. *Habitat.* Found in the vicinity of fresh water and coastal lagoons. Common in most parts but absent from large tracts in the central, western and southern regions. *Call.* A loud, shrill 'chee-chee-cheree-cheeroo'. *Afrikaans.* Bontkwikkie.

713 **Cape Wagtail** *Motacilla capensis* L 18 cm

The unmarked, greyish brown upperparts, combined with the narrow black breast band, are diagnostic. Distinguished from the Longtailed Wagtail by having a shorter tail, buffy, not white belly and flanks, and by lacking extensive white in the wings. *Habitat.* Usually near fresh water or coastal lagoons but has also adapted to city parks and gardens. Common throughout except in the north-east. *Call.* A clear, ringing 'tseee-chee-chee' call and a whistled, trilling song. *Afrikaans.* Gewone Kwikkie.

712 **Longtailed Wagtail** *Motacilla clara* L 20 cm

Its unusually long tail, pale grey upperparts and white, not buff underparts distinguish this species from the Cape Wagtail. It also shows much more white in the wings and is altogether a far more slender bird. *Habitat.* Confined to fast-flowing streams in evergreen and coastal forests. Common but localized in scattered eastern regions. *Call.* A sharp, high-pitched 'cheeerip' or 'chissik'. *Afrikaans.* Bergkwikkie.

714 **Yellow Wagtail** *Motacilla flava* L 16 cm

Several races occur within the region. However, females and the non-breeding males which are normally seen in our region, are indistinguishable in the field and show a wide variation in the amount of yellow on underparts. Smaller than the Grey Wagtail, this species has a much shorter tail, and a green, not blue-grey back. *Habitat.* Grassy surrounds of coastal lagoons, sewage ponds and freshwater areas. Uncommon, but sometimes locally common in the region except for the dry central and western areas. A summer visitor from Eurasia. *Call.* A weak, thin 'tseeep'. *Afrikaans.* Geelkwikkie.

715 **Grey Wagtail** *Motacilla cinerea* L 18 cm

This species might be mistaken for the Yellow Wagtail but has a blue-grey, not green back which contrasts with its greenish yellow rump. The flanks and vent are a bright sulphur yellow. It also has a noticeably longer tail. *Habitat.* Near fast-flowing streams and in grassy areas adjacent to fresh water. Rare vagrant to Natal, the Transvaal and Namibia. *Call.* A single, sharp 'tit'. *Afrikaans.* Gryskwikkie.

716 Richard's Pipit *Anthus novaeseelandiae* L 16 cm
Difficult to distinguish from the Longbilled Pipit but generally shows bolder facial markings and breast streaking and, in flight, displays white, not buff outer tail feathers. Differs from Mountain Pipit by its call, the white outer tail feathers and the yellow, not pink base to its lower mandible. *Habitat.* Virtually any type of open grassland. Common throughout the region. *Call.* Song given during display flight is a 'trrit-trrit-trrit'. When flushed it utters a 'chisseet'. *Afrikaans.* Gewone Koester.

717 Longbilled Pipit (Nicholson's Pipit) *Anthus similis* L 18 cm
The only reliable field differences between this and Richard's Pipit are this species' buff, not white outer tail feathers, less distinct facial markings and streaking on breast, and the song and call. Difficult to distinguish from Mountain Pipit but has a less boldly streaked breast and has a different call and display. *Habitat.* Usually prefers boulder-strewn hillsides with scant bush cover, but does occur in broad-leafed woodland in the north. Common throughout the region in scattered localities. Absent from coastal lowlands in the east. *Call.* A high-pitched trisyllabic 'pheeet-pheeet-cher' (unlike the duosyllabic call of the Plainbacked Pipit) and a sparrow-like 'cheerup'. *Afrikaans.* Nicholsonse Koester.

718 Plainbacked Pipit *Anthus leucophrys* L 18 cm
Differs from Richard's and Longbilled pipits by having a uniform, unstreaked back, by lacking distinct breast markings and by its narrow, buff outer tail feathers. Very difficult to distinguish from the Buffy Pipit except by call and song although it has generally darker upperparts and there is far less contrast between its buffy flanks and belly. *Habitat.* Hillsides covered with short grass and, when not breeding, this species flocks to stubble grain fields. Common, but erratically distributed in the south, east and extreme north. *Call.* A loud, clear duosyllabic 'chrrrup-chereeoo', similar in quality to the Longbilled Pipit's trisyllabic call. *Afrikaans.* Donkerkoester.

719 Buffy Pipit *Anthus vaalensis* L 18 cm
A species which varies in plumage coloration, it is very difficult to distinguish from the Plainbacked Pipit. Main differences lie in the very faintly dappled, not streaked breast, contrasting with the rich, warm buffy belly and flanks. On the ground it behaves more like a wagtail, stopping often, and slowly moving its tail up and down. *Habitat.* Similar to that frequented by the Plainbacked Pipit (hillsides covered with short grass) but usually at lower altitudes. Flocks visit burnt grassy areas. Common in the northern and central regions.
Call. When flushed gives a short 'sshik'. Song, a 'tchipeep-cheree', is softer than that of the Plainbacked Pipit. *Afrikaans.* Vaalkoester.

901 Mountain Pipit *Anthus* (specific name not yet decided) L 18 cm
Difficult to distinguish from Richard's Pipit with which it often associates, but is larger, has very bold streaking on the breast, a bright pink, not yellow base to the lower mandible and, in flight, shows buff, not white outer tail feathers. *Habitat.* Montane grasslands, usually at an altitude above 2 000 m. Common summer visitor to the north-eastern Cape, Lesotho and possibly Natal. *Call.* Almost indistinguishable from that of Richard's Pipit but is slightly deeper in pitch and slower in tempo.

901▲

719▲

718▲   716▼ 717▼

721 Rock Pipit *Anthus crenatus* L 17 cm
A drab, uniformly coloured pipit with a stout, heavy bill. Only at close range can the faint, narrow streaking on the breast and the greenish edges to the secondary coverts, be seen. Usually located by its distinctive song. *Habitat.* Boulder- and rock-strewn steep, grassy hillsides. A common but often overlooked endemic to the south and south-east. *Call.* A carrying 'tsseeee-tcherrrrooo'. *Afrikaans.* Klipkoester.

724 Short-tailed Pipit *Anthus brachyurus* L 12 cm
A small, squat pipit, very much darker and more heavily streaked above and below than either the Tree or Bushveld Pipit. In flight, shows a noticeably shorter, thinner tail than other small pipits. When flushed, flies off speedily, resembling a large cisticola. *Habitat.* Grassy hillsides, scantily covered with protea scrub, and grassy glades in miombo woodland. *Call.* Similar to Bushveld Pipit, a buzzy 'chrrrrt-zhrrrreet-zzeeep'. *Afrikaans.* Kortstertkoester.

722 Tree Pipit *Anthus trivialis* L 15 cm
Longer-tailed and larger than either the Bushveld or Short-tailed Pipit and has a short, dark bill. Differentiated with difficulty from the Redthroated Pipit but has less clearly streaked underparts and lacks any dark brown streaking on its rump. *Habitat.* Grassy areas in open broad-leafed woodland. Uncommon summer visitor to the north. *Call.* A soft, nasal 'teeez'. Song not recorded in our region. *Afrikaans.* Boomkoester.

723 Bushveld Pipit *Anthus caffer* L 13,5 cm
Distinguished from the Tree Pipit by being smaller, having a shorter tail, much paler plumage, and suffused, less distinct streaking on its breast. In comparison with the Short-tailed Pipit, it is larger, has a longer, broader tail and is less heavily streaked below. *Habitat.* Thornveld and open broad-leafed woodland. Uncommon and localized in the eastern and north-central regions. *Call.* A treble call note: 'zrrrt-zrree-chreee'. *Afrikaans.* Bosveldkoester.

720 Striped Pipit *Anthus lineiventris* L 18 cm
A very plump, heavily built pipit which differs from all others by its boldly striped underparts, dark upperparts with yellow-edged wing feathers, and very dark tail with conspicuous white outer tail feathers. Often seen running along the thicker branches of trees. *Habitat.* Thickly wooded, boulder-strewn hill slopes. Common, but highly localized, in the south-eastern and north-central regions. *Call.* An extremely loud, penetrating thrush-like song. *Afrikaans.* Gestreepte Koester.

903 Redthroated Pipit *Anthus cervinus* L 14 cm
Noticeably smaller than Richard's Pipit. Most likely to be confused with Tree Pipit from which it differs by being generally darker, having clear white underparts heavily streaked with black, and a streaked, not uniformly coloured rump. *Habitat.* Damp grasslands, usually near water. Rare Palearctic vagrant, with one record from Natal. *Call.* A clear, penetrating 'chup', and a buzzy 'skeeeaz'.

274

721 ▲   722 ▼   723 ▼   724 ▲

903 ▼   720 ▼

725 Yellowbreasted Pipit *Anthus chloris* L 17 cm
The only pipit in the region to show an all-yellow throat and breast. The upperparts, seen when the bird runs through grass, are heavily streaked and blotched, giving a scaled appearance, and are similar to those of a longclaw. In flight, the yellow underparts and wing linings are diagnostic. *Habitat.* Long and short grass tufts on grassy mountain slopes, usually at an altitude above 1 500 m. Uncommon endemic to the higher regions of the central south-east. *Call.* A long, buzzy 'trzzzzzzit' note. *Afrikaans.* Geelborskoester.

726 Golden Pipit *Tmetothylacus tenellus* L 15 cm
Although it resembles a diminutive, bright golden Yellowthroated Longclaw, the predominantly vivid yellow flight feathers render this bird unmistakable. The female is slightly duller than the male and lacks the black breast band. *Habitat.* Open, dry broad-leafed woodland and thornveld. An extremely rare vagrant from tropical Africa, with records from Zimbabwe and the Transvaal. *Call.* Not recorded in our region. *Afrikaans.* Goudkoester.

727 Orangethroated Longclaw *Macronyx capensis* L 20 cm
Could be confused with the Yellowthroated Longclaw but has deeper yellow underparts and a diagnostic orange throat encircled with black. Imm. differs from imm. Yellowthroated by having buffish to orange underparts, and wing feathers edged with buff, not yellow. *Habitat.* Wide range of coastal and upland grasslands. A common and widespread endemic to the southern and eastern sectors. *Call.* A cat-like mewing and a loud, sparrow-like 'cheeerrup'. *Afrikaans.* Oranjekeelkalkoentjie.

728 Yellowthroated Longclaw *Macronyx croceus* L 20 cm
Easily distinguished from the Orangethroated Longclaw by having bright yellow underparts and a yellow, not orange throat patch. Imm. differentiated from imm. Orangethroated by being buff-yellow below and by having bright yellow edging to its wing feathers. Both species are plump, short-tailed birds which, when flushed, show conspicuous white outer tail feathers. *Habitat.* Grasslands adjoining freshwater areas, coastal estuaries and lagoons. Common in the eastern lowlands. *Call.* A loud, whistled 'phooooeeet'. *Afrikaans.* Geelkeelkalkoentjie.

730 Pinkthroated Longclaw *Macronyx ameliae* L 20 cm
Much more pipit-like in shape than either the Yellow- or Orangethroated Longclaw, being slender and long-tailed. The bright pink throat and breast are diagnostic. Imm. lacks the pink but shows a rosy hue over its belly and flanks. *Habitat.* Moist grasslands surrounding open areas of fresh water. Uncommon and thinly distributed in the north-east and in northern Botswana. *Call.* A sharp, pipit-like 'chiteeet'. *Afrikaans.* Rooskeelkalkoentjie.

729 Fülleborn's Longclaw *Macronyx fuellebornii* L 20 cm □
Although there are no positively confirmed records from our region, this species may be found just south of the Angola border, especially in damp, grassy areas. It resembles a dull, faded version of the Yellowthroated Longclaw. *Afrikaans.* Angolakalkoentjie.

276

725 ▲   726 ◢

730 ▲   728 ▼   727 ▼

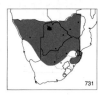

731 **Lesser Grey Shrike** *Lanius minor* L 20 cm
In comparison with the Redbacked Shrike, this species is larger, has a more extensive black mask which encompasses the forehead, and a grey, not chestnut back. Imm. differs from the imm. and female Redbacked Shrikes by its much larger size and buffier, less barred underparts. *Habitat.* Mixed dry thornveld and semi-desert scrub. Common summer visitor from Eurasia to the northern sector. *Call.* A soft 'chuk'. Song seldom heard in our region. *Afrikaans.* Gryslaksman.

732 **Fiscal Shrike** *Lanius collaris* L 23 cm
The similar Boubou Shrikes are shy, skulking birds whereas this species is one of the most familiar shrikes of the region, hunting from exposed perches along roadsides and in suburbia. Ad. male is black above, white below with prominent white shoulder patches. Female shows a rust-coloured patch on the flanks. Imm. is greyish brown with grey crescentic barring below. Ad. differs from male Fiscal Flycatcher by lacking white sides to the tail and by having a hooked, not flattened bill. *Habitat.* Found in virtually every habitat except dense forest. Common throughout except in the north-central and eastern regions. *Call.* Harsh grating, a melodious whistled song jumbled with harsher notes, and mimicry of other bird calls. *Afrikaans.* Fiskaallaksman.

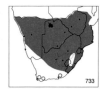

733 **Redbacked Shrike** *Lanius collurio* L 17 cm
In the region, the only shrike with a chestnut-coloured back. The male's blue-grey head and rump contrast with the chestnut back and are diagnostic. Female and imm. are reddish brown above and have greyish brown crescentic barring below. *Habitat.* Frequents a wide variety of broad-leafed woodland and thornveld. Common summer visitor from Eurasia to the greater part of the region, but absent from the extreme south. *Call.* A harsh 'chak, chak' and a soft, warbler-like song. *Afrikaans.* Rooiruglaksman.

734 **Sousa's Shrike** *Lanius souzae* L 18 cm
Resembles a very pale female Redbacked Shrike but is easily distinguished by its white shoulder patches and very pale grey head. Distinguished from imm. Fiscal Shrike by its smaller size and by the lack of crescentic barring below. *Habitat.* Miombo and mopane woodlands. Vagrant to the extreme north-central region. *Call.* Not recorded in our region. *Afrikaans.* Souzase Laksman.

735 **Longtailed Shrike** *Corvinella melanoleuca* L 40-50 cm
The combination of the black and white plumage and the long, wispy tail is diagnostic and unmistakable. The female has a shorter, white-tipped tail and has whitish flanks. *Habitat.* Thornveld and open broad-leafed woodland. Common in the northern and central parts of the region. *Call.* A liquid, whistled 'peeleeo'. *Afrikaans.* Langstertlaksman.

739 **Crimsonbreasted Shrike** *Laniarius atrococcineus* L 23 cm
This bird has a striking combination of bright crimson underparts and black upperparts which render it unmistakable. A rare, aberrant form occurs in which the crimson is replaced with yellow. Imm. is grey below; above, it is finely barred black with buff edges to the feathers. *Habitat.* Thornveld and dry rivercourses. Common endemic to the dry central and western regions. *Call.* A harsh 'trrrrr' and a whistled 'pheeee-tcherooo' duet. *Afrikaans.* Rooiborslaksman.

278

733 ▲   735 ▼   734 ▼   732 ▲

731 ▼   739 ▼

736 Southern Boubou *Laniarius ferrugineus* L 23 cm
Similar to the Fiscal Shrike but has a shorter tail, has the white in the wing extending on to the secondaries, and is less bold in its habits. Distinguished from Tropical and Swamp boubous by having rust-coloured flanks, undertail and belly. *Habitat.* Creeps through thickets in riverine and evergreen forests. Common endemic to the east and extreme south. *Call.* A variable duet with basic notes of 'boo-boo' followed by 'whee-ooo'. *Afrikaans.* Suidelike Waterfiskaal.

737 Tropical Boubou *Laniarius aethiopicus* L 21 cm
Easily confused with the Southern Boubou but is generally paler below, showing a more marked contrast between its black upperparts and pinkish white underparts. The pink-tinged underparts also distinguish this species from the Swamp Boubou which is pure white below. *Habitat.* Thickets in riverine and evergreen forests. Common in the north-east, where it replaces the Southern Boubou. *Call.* Duet call, very similar to that of the Southern Boubou. *Afrikaans.* Tropiese Waterfiskaal.

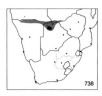

738 Swamp Boubou *Laniarius bicolor* L 22 cm
Distinguished from both the Southern and Tropical boubous by having clear white underparts with no trace of rust or pink coloration. *Habitat.* Thickets alongside rivers, and papyrus swamps. Common in northern Botswana. *Call.* A clear, bell-like 'tuuwhooo'. *Afrikaans.* Moeraswaterfiskaal.

742 Southern Tchagra *Tchagra tchagra* L 22 cm
Differs from the similar Threestreaked Tchagra by having a longer, heavier bill and by lacking conspicuous black stripes bordering its buff eyebrow stripe. It is also slightly larger and is darker in appearance. *Habitat.* Coastal scrub, forest edges and thickets. Common endemic to the coastal south-east and south. *Call.* Song given in aerial display is a 'wee-chee-chee-cheee'. *Afrikaans.* Grysborstjagra.

743 Threestreaked Tchagra *Tchagra australis* L 18 cm
Very similar in appearance to the Southern Tchagra, this species has a paler crown and forehead, and its broad white eyebrow stripe is bordered by black stripes. It is also smaller and paler, and has a less massive bill. *Habitat.* Thick tangles and undergrowth in thornveld. Common throughout the north of the region. *Call.* Aerial display flight and song are very similar to that of the Southern Tchagra. *Afrikaans.* Rooivlerktjagra.

744 Blackcrowned Tchagra *Tchagra senegala* L 23 cm
The black forehead and crown, and paler underparts distinguish this species from other tchagras. It is also bolder and more conspicuous in its behaviour. *Habitat.* Mixed thornveld and riverine scrub. Common in the north and south-east. *Call.* Song is a loud, whistled 'whee-cheree, cherooo, cheree-cherooo' on a descending scale, slurring off towards the end. *Afrikaans.* Swartkroontjagra.

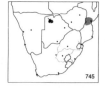

745 Marsh Tchagra (Blackcap Tchagra) *Tchagra minuta* L 18 cm
This marsh-dwelling shrike is easily identified by its black cap, russet upperparts and creamy to buffish underparts. The female lacks the black cap and has a broad, buff eyebrow stripe. Imm. is unusual in that it has a buff-white crown. Very shy and furtive, not easily located unless heard calling or singing. *Habitat.* Rank bracken and briar growing in damp hollows, and marshy areas with long grass. Uncommon and localized in Moçambique and Zimbabwe. *Call.* A shrill, trilling song is given in display flight. *Afrikaans.* Vleitjagra (Swartkoptjagra).

280

737 ▲

736 ▲

743 ▲

738 ▲   742 ▼ 745 ▼   744 ▲

746 Bokmakierie *Telophorus zeylonus* L 23 cm
The bright lemon yellow underparts and the broad black breast band are diagnostic. The vivid yellow tip to the dark green tail is conspicuous when the bird is in flight or diving into bushes for cover. *Habitat.* Fynbos, Karoo scrub and scrub-filled valleys in mountainous terrain. Common endemic to the south and west, and the highlands on the border of Zimbabwe and Moçambique. *Call.* Very varied, but is usually a 'bok-bok-kik'. *Afrikaans.* Bokmakierie.

747 Gorgeous Bush Shrike *Telophorus quadricolor* L 20 cm
More often heard than seen but when glimpsed it is unmistakable – it has a bright red throat, black breast band and yellow-orange belly. Imm. has a yellow throat, lacks the black breast band, and is distinguished from imm. Orangebreasted and Blackfronted bush shrikes by having green, not grey upperparts. *Habitat.* Thick coastal and riverine forests. Common in the east and north-east. *Call.* An often-repeated 'conk-conk-queet' and variations of this call. *Afrikaans.* Konkoit.

751 Greyheaded Bush Shrike *Malaconotus blanchoti* L 26 cm
The size of this, the largest bush shrike of the region, and its heavy, hooked bill eliminate confusion with any similarly coloured shrike. The bright yellow underparts sometimes have a faint orange wash across the breast. *Habitat.* Thornveld and mixed broad-leafed woodland. Common, but easily overlooked, in the eastern sector of the region. *Call.* A drawn-out 'oooooop' and a 'tic-tic-oooop'. *Afrikaans.* Spookvoël.

748 Orangebreasted Bush Shrike *Telophorus sulfureopectus* L 19 cm
Differs from all other similar bush shrikes, especially the Olive, by having a conspicuous yellow forehead and eyebrow stripe. Female is duller than the male, and the imm. is distinguished from the imm. Blackfronted Bush Shrike by being paler grey on the head and paler yellow below. *Habitat.* Thornveld and riverine forest. Common in the north and east. *Call.* Frequently repeated 'poo-poo-poo-poooo' and a 'titit-eeezz'. *Afrikaans.* Oranjeborsboslaksman.

749 Blackfronted Bush Shrike *Telophorus nigrifrons* L 19 cm
The dark blue-grey crown and nape do not contrast markedly with the black mask and this gives the bird a dark-capped appearance. Although the cap and bright orange underparts render this shrike readily identifiable, it does superficially resemble the Chorister Robin. *Habitat.* A less skulking species than other bush shrikes, it frequents the canopy of evergreen forests, and verges. Uncommon and highly localized in the north-east. *Call.* A harsh 'tic-chrrrrr' and a ringing 'oo-pooo' call. *Afrikaans.* Swartoogboslaksman.

750 Olive Bush Shrike *Telophorus olivaceus* L 18 cm
Occurs in two widely differing colour phases. The ruddy phase has a blue-grey head and, unlike any other bush shrike, has a white eyebrow stripe. The olive phase is duller: the bird has a green, not blue-grey head and it lacks the white eyebrow stripe. *Habitat.* Evergreen and riverine forests. A common but localized endemic to the south and east, with a small population in Malawi. *Call.* A whistled 'cheeoo-cheeoo-cheeoo-cheeoo' and a call similar to the Orangebreasted Bush Shrike's 'poo-poo-poo-poooo'. *Afrikaans.* Olyfboslaksman.

746 ▲

747 ▲

748 ▼

750 ▲

751 ▲

749 ▼

753 White Helmetshrike *Prionops plumatus* L 20 cm
The grey crown, white collar, white flashes in wings and the white outer tail feathers are diagnostic. The birds usually gather in small groups and follow each other through woodland. *Habitat.* Mixed woodlands and thornveld, avoiding thick evergreen forests. Common and conspicuous in the north and east. *Call.* A jumble of chatters and whistles. *Afrikaans.* Withelmlaksman.

754 Redbilled Helmetshrike *Prionops retzii* L 22 cm
Differs from the Chestnutfronted Helmetshrike by lacking the chestnut forehead and by having jet black, not dark grey plumage. At close range the bright red eye-ring, bill and legs are noticeable. *Habitat.* Mostly mopane and miombo woodland. Common in the north and north-east. *Call.* Loud whistles and harsh chattering. *Afrikaans.* Swarthelmlaksman.

755 Chestnutfronted Helmetshrike *Prionops scopifrons* L 22 cm
The chestnut-coloured forehead distinguishes this species from the Redbilled Helmetshrike. It also differs by having a dark grey, not jet black body. *Habitat.* Coastal and riverine forests. Uncommon and localized in the north-east. *Call.* Not recorded in our region. *Afrikaans.* Stekelkophelmlaksman.

756 Whitecrowned Shrike *Eurocephalus anguitimens* L 25 cm
This large, robust bird is the only shrike in the region to have a white crown and forehead. The throat and breast are also white, while the belly and flanks are washed with buff. *Habitat.* Mixed, dry woodland and thornveld. Common endemic to the north and north-west. *Call.* A shrill, whistling 'kree, kree, kree' and harsh chattering. *Afrikaans.* Kremetartlaksman.

741 Brubru *Nilaus afer* L 15 cm
This bird has all the appearances of a large batis but its size and thick bill should obviate any confusion. The black and white chequered back, broad white eyebrow stripe and russet flank stripe are diagnostic. *Habitat.* Dry thornveld and open broad-leafed woodland. Common in suitable habitat throughout the region except the extreme south. *Call.* A soft, trilling 'prrrrr, prrrrr' and a piercing whistle 'tioooo'. *Afrikaans.* Bontroklaksman.

740 Puffback *Dryoscopus cubla* L 18 cm
The only shrike in the region to have a large white rump which is conspicuous when it is spread and puffed up during display. At other times, the bird resembles a small boubou but is shorter tailed, has white-scaled wing coverts and has a bright red eye. *Habitat.* Frequents a wide variety of woodland and forest. Common in the wetter parts of the region, being absent from the dry central and western areas. *Call.* A whistled 'weeee', followed by a whip-like 'cheraaak'. *Afrikaans.* Sneeubal.

752 Whitetailed Shrike *Lanioturdus torquatus* L 15 cm
The striking black, white and grey plumage, long legs and very short, all-white tail are diagnostic. Often seen on the ground or hopping over rocks and, with its very upright posture, it appears almost tailless. *Habitat.* Dry thornveld and scrub desert. Common endemic to Namibia. *Call.* Clear, drawn-out whistles and harsh cackling. *Afrikaans.* Kortstertlaksman.

753 ▲

752 ▼

741 ▲

755 ▲

756 ▲

740 ▼

754 ▼

764 Glossy Starling *Lamprotornis nitens* L 25 cm
The short-tailed 'glossy' starlings are difficult to distinguish in the field unless seen at close range. This species differs from the Greater and Lesser Blue-eared starlings by having uniform glossy green ear patches and head, and glossy green belly and flanks. Distinguished from Blackbellied Starling by its generally brighter, shinier plumage and by lacking the dull black belly. *Habitat.* Thornveld, mixed woodland and suburbia. Common and widespread throughout the region except the extreme south. *Call.* Song is a slurred 'trrr-chree-chrrrr'. *Afrikaans.* Kleinglansspreeu.

765 Greater Blue-eared Starling *Lamprotornis chalybaeus* L 23 cm
Distinguished from the Glossy Starling by having a broad dark blue, not green ear patch, and blue, not green belly and flanks. Larger than the Lesser Blue-eared Starling, it also has a broader ear patch, and blue, not magenta belly and flanks. *Habitat.* Thornveld and mopane woodland. Common in the north-east. *Call.* A nasal 'squee-aar' and a warbled song. *Afrikaans.* Groot-blouoorglansspreeu.

766 Lesser Blue-eared Starling *Lamprotornis chloropterus* L 20 cm
Although it could be confused with the Greater Blue-eared Starling, this is a smaller bird with a more compact head and a finer bill. The dark blue ear patch is less extensive and appears as a black line through and behind the eye. The belly and flanks are magenta, not blue. *Habitat.* Confined to miombo woodland. Common but very localized in the north-east. *Call.* Higher pitched than that of the Greater Blue-eared Starling, and with a 'wirri-girri' flight call. *Afrikaans.* Klein-blouoorglansspreeu.

768 Blackbellied Starling *Lamprotornis corruscus* L 21 cm
The least glossy of all the starlings in this group, it shows a black belly and flanks. Imm. and female are duller than the male and appear black in the field. *Habitat.* Common in coastal and riverine forests along the east coast. *Call.* Harsh, chippering notes interspersed with shrill whistles. *Afrikaans.* Swartpensglansspreeu.

761 Plumcoloured Starling *Cinnyricinclus leucogaster* L 19 cm
The male's upperparts and throat are an unusual gaudy, glossy amethyst colour. Female is mottled and streaked dark brown above, and the white underparts are heavily streaked with brown. *Habitat.* Found in most woodlands but avoids thick, evergreen forests. Common summer visitor from tropical Africa to the northern and eastern sectors of the region. *Call.* A soft, but sharp 'tip, tip'. Song is a short series of buzzy whistles. *Afrikaans.* Witborsspreeu.

764 ▲

765 ▼

768 ▼   766 ▲

761 ▼

763 Longtailed Starling *Lamprotornis mevesii* L 34 cm
The long, graduated and pointed tail is diagnostic. Although similar, this species is a smaller, more compact bird than Burchell's Starling. Imm. is duller in colour than ad. but the long, pointed tail is evident. *Habitat.* Tall, mopane woodland and riverine forest. Common but localized in the north. *Call.* A harsh 'keeeaaaa' and churring notes. *Afrikaans.* Langstertglansspreeu.

767 Sharptailed Starling *Lamprotornis acuticaudus* L 26 cm
Difficult to distinguish from the other short-tailed glossy starlings unless the diagnostic wedge-shaped tail is seen. In flight, the underside of the primaries appears pale, not black as in similar starlings. *Habitat.* Dry, broad-leafed woodland and dry rivercourses. Uncommon in northern Namibia and Botswana. *Call.* Not yet recorded in our region.
Afrikaans. Spitsstertglansspreeu.

762 Burchell's Starling *Lamprotornis australis* L 34 cm
A large, glossy starling which appears much more heavily built than the similar Longtailed Starling. This species has broader, more rounded wings and a shorter, broader tail. *Habitat.* Thornveld and dry, broad-leafed woodland. Common endemic to the central and north-western regions. *Call.* Song is a jumble of throaty chortles and chuckles. *Afrikaans.* Grootglansspreeu.

770 Palewinged Starling *Onychognathus nabouroup* L 26 cm
Liable to be confused with the Redwinged Starling, but differs by having a white, not chestnut patch in the primaries. When at rest, this white patch appears to be orange-edged, not entirely bright chestnut as in the Redwinged. Shows a red, not dark brown eye. *Habitat.* Rocky ravines and cliffs in dry and desert regions. Common in the dry west, where it replaces the Redwinged Starling. *Call.* Very similar to that of the Redwinged Starling. *Afrikaans.* Bleekvlerkspreeu.

769 Redwinged Starling *Onychognathus morio* L 27 cm
Bright chestnut flight feathers and dark brown, not red eyes distinguish this species from the Palewinged Starling. The female has an ash grey head and upper breast. *Habitat.* Rocky ravines, cliffs, and has adapted to suburbia. Common in the eastern and northern sectors. *Call.* A clear, whistled 'cherleeeeoo' and a variety of other whistles. *Afrikaans.* Rooivlerkspreeu.

767▲

763▲

770▼

769▼

762▲

757 European Starling *Sturnus vulgaris* L 21 cm
Ad. is easily identified by its yellow beak and glossy black plumage speckled with white. Imm. is very similar to the female Wattled Starling but shows an obvious pale grey throat and lacks the pale rump. *Habitat.* Frequents a wide range of habitats, from cities to open farmland. Common in the southern Cape. *Call.* Song includes mimicry, whistles and chattering. *Afrikaans.* Europese Spreeu.

759 Pied Starling *Spreo bicolor* L 27 cm
This large, dark brown starling is distinguished by its conspicuous white vent and undertail coverts. At close range the diagnostic yellow base to the lower mandible, and the bright lemon eyes can be seen. *Habitat.* A commonly seen roadside bird, it is also found in grasslands. Common endemic to the southern part of the region. *Call.* A loud 'skeer-kerrra-kerrra'. *Afrikaans.* Witgatspreeu.

760 Wattled Starling *Creatophora cinerea* L 21 cm
The breeding male is distinctive with its pale grey body, black and yellow head and black wattles. The non-breeding male and female are nondescript grey birds, which show black wings and tails, and contrasting pale grey rumps. *Habitat.* Grasslands and open broad-leafed woodlands. Common but nomadic throughout the region. *Call.* Various hisses and cackles, and a 'ssreeeeo' note. *Afrikaans.* Lelspreeu.

758 Indian Myna *Acridotheres tristis* L 25 cm
The only myna of the region and unlikely to be confused with any starling because of its white wing patches, white tips to tail and bare yellow skin around the eyes. Moulting adults sometimes lose most of their head feathers, and as a result, the head then appears yellow. *Habitat.* Urban and suburban regions. Common throughout Natal and in Johannesburg. *Call.* Jumbled titters and chattering. *Afrikaans.* Indiese Spreeu.

771 Yellowbilled Oxpecker *Buphagus africanus* L 22 cm
A much paler bird than the Redbilled Oxpecker and easily identified by its bright yellow bill with a red tip, and pale lower back and rump. Imm. has a brown, not black bill and is much paler than the imm. Redbilled Oxpecker. *Habitat.* Thornveld and broad-leafed woodland, often near water. Frequently found in association with buffalo, rhino and hippo. Locally common in the north. *Call.* A short, hissing 'kriss, kriss'. *Afrikaans.* Geelbekrenostervoël.

772 Redbilled Oxpecker *Buphagus erythrorhynchus* L 22 cm
Distinguished from the Yellowbilled Oxpecker by having an all-red bill, bare yellow skin around its eyes and a dark, not pale rump. Imm. has a yellow base to a black bill and is darker than the imm. Yellowbilled Oxpecker, showing a dark, not pale rump. *Habitat.* Thornveld and broad-leafed woodland. Common in the north and east. Prefers to associate with hairier mammals such as giraffe and buck. *Call.* A scolding 'churrrr' and a hissing 'zzzzzzist'. *Afrikaans.* Rooibekrenostervoël.

757 ▲   759 ▲

771 ▲   760 ▼   758 ▼   772 ▲

773 Cape Sugarbird *Promerops cafer* L ♂ 37-44 cm; ♀ 25-29 cm
Confusion between this species and Gurney's Sugarbird should not occur as
their distribution ranges are mutually exclusive (except for a localized area in
the eastern Cape), and this species is easily recognized by its exceptionally
long, wispy tail. Lacks the russet breast and crown of Gurney's Sugarbird and
has distinct malar stripes. *Habitat.* Stands of flowering proteas on mountain
slopes, and in commercial protea nurseries. A common endemic to the
southern Cape. *Call.* Starling-like chirps and whistles. .
Afrikaans. Kaapse Suikervoël.

774 Gurney's Sugarbird *Promerops gurneyi* L 23-29 cm
Unlikely to be confused with the Cape Sugarbird because of its smaller size,
shorter tail, conspicuous russet breast and crown, and lack of malar stripes.
Habitat. Stands of flowering proteas and aloes in mountainous regions.
A common but localized endemic to eastern mountain ranges. *Call.* A higher
pitched, more melodious rattling song than that of the Cape Sugarbird.
Afrikaans. Rooiborssuikeryoël.

775 Malachite Sunbird *Nectarinia famosa* L ♂ 25 cm; ♀ 15 cm
The only sunbird in the region to have an entirely metallic green plumage and
long, pointed tail projections. The Bronze Sunbird also has tail projections but
appears black in the field. The female is very similar to the female Bronze
Sunbird but is larger, paler yellow below and lacks distinct streaking. In non-
breeding and eclipse plumage, the male loses most of its metallic green
coloration and becomes drab brown above, like the female. *Habitat.* Fynbos,
protea- and aloe-covered hills, and mountain scrub. Common in the south but
rarer further north and east. *Call.* A ringing 'zing-zing'. Song is a series of
chipping notes. *Afrikaans.* Jangroentjie.

776 Bronze Sunbird *Nectarinia kilimensis* L ♂ 21 cm; ♀ 14 cm
Smaller than the similarly shaped Malachite Sunbird. In the field, unless the
bird is seen in direct sunlight, the metallic bronze plumage of the male
appears black. The female is very similar to the female Malachite Sunbird but
is smaller, has a shorter, more decurved bill and has brighter yellow
underparts which are distinctly streaked with brown. *Habitat.* Evergreen forest
edges and adjoining grasslands. Common but localized in the highlands of
Zimbabwe and Moçambique. *Call.* A loud, piercing 'chee-oo'.
Afrikaans. Bronssuikerbekkie.

777 Orangebreasted Sunbird *Nectarinia violacea* L ♂ 15 cm; ♀ 12 cm
The only small sunbird in the region to have pointed tail projections. The
male's dark head and throat, and orange-yellow belly and breast are
diagnostic. The female is larger than the female Lesser Doublecollared
Sunbird and is a more uniform olive green above and below. *Habitat.* Fynbos,
and flowering montane protea and aloe stands. Common endemic to the
extreme southern Cape. *Call.* A harsh, buzzy 'tsee-aap' and a 'teer-turp'.
Afrikaans. Oranjeborssuikerbekkie.

775▲   777▼ 776▼   774▼   773▲

792 Black Sunbird *Nectarinia amethystina* L 15 cm
The only all-black sunbird of the region, but at close range a greenish bronze iridescence can be seen on the throat and forehead. The female could be confused with the female Scarletchested Sunbird but has paler, more uniform upperparts, less heavily streaked underparts, and obvious buff moustachial stripes. *Habitat.* Evergreen, coastal and riverine forests, and suburbia. Common in the eastern sector of the region. *Call.* A fast, twittering song. *Afrikaans.* Swartsuikerbekkie.

778 Coppery Sunbird *Nectarinia cuprea* L 12 cm
Resembles a small Bronze Sunbird but lacks the pointed tail projections of that species. The female, which is far smaller than the female Malachite and Bronze sunbirds, is a much brighter yellow below with faint speckling on the throat and breast. *Habitat.* Diverse in choice of habitat but is usually found in woodland, forest clearings and edges, and moist riverine forest. Common, but very localized, in the extreme north-east. *Call.* A harsh 'chit-chat', and a soft warbling song. *Afrikaans.* Kopersuikerbekkie.

791 Scarletchested Sunbird *Nectarinia senegalensis* L 15 cm
The male's black body and scarlet chest are diagnostic. Female is distinguished from the female Black Sunbird by the very heavily mottled underparts and by lacking the buff moustachial stripes. *Habitat.* Mixed dry and moist woodland, thornveld and suburbia. Common and widespread in the north and south-east. *Call.* A loud, whistled 'cheeup, chup, toop, toop, toop' song. *Afrikaans.* Rooikeelsuikerbekkie.

790 Olive Sunbird *Nectarinia olivacea* L 14 cm
Likely to be confused only with the Grey Sunbird which occurs in the same habitat, but this species has very obvious olive, not grey coloration and yellow, not red pectoral tufts. It also has a different call. *Habitat.* Coastal and riverine forest and moist, broad-leafed woodland. Common in the east and south-east. *Call.* A sharp 'tuk, tuk, tuk' and a short, bouncy 'cheef-chaf-cheef' song. *Afrikaans.* Olyfsuikerbekkie.

789 Grey Sunbird *Nectarinia veroxii* L 14 cm
A fairly nondescript sunbird which has uniformly grey underparts and darker grey upperparts. Differs from the Olive Sunbird in its call, by its grey, not olive coloration and by having red, not yellow pectoral tufts. *Habitat.* Coastal and riverine forest. Common but localized along the eastern and south-eastern coast. *Call.* A harsh, grating 'tzzik, tzzik' and a short 'chrep, chreep, peepy' song. *Afrikaans.* Gryssuikerbekkie.

781 Shelley's Sunbird *Nectarinia shelleyi* L 12 cm
Although it occurs in association with the very similar Miombo Doublecollared Sunbird, it can be distinguished by its black, not pale olive belly and green, not blue rump. Female differs from the female Miombo Doublecollared Sunbird by having the yellow underparts lightly streaked with brown. This species is similar to Neergaard's Sunbird but their ranges are mutually exclusive. *Habitat.* Miombo and moist riverine scrub. Doubtful if it occurs south of the Zambezi River in Zimbabwe. Recently reported from northern Botswana. *Call.* Described as a 'didi-didi', and a nasal 'chibbee-cheeu-cheeu' song. *Afrikaans.* Swartpenssuikerbekkie.

778▲   791▼ 790▼   781▲

792▼ 789▼

788 Dusky Sunbird *Nectarinia fusca* L 10 cm
The slightly metallic black head and throat and contrasting white belly are diagnostic in the male. However, male plumage is variable and sometimes shows only a black line running from the throat on to the breast. Female resembles the female Lesser Doublecollared Sunbird but is much paler and has completely off-white underparts. *Habitat.* Dry thornveld, dry, wooded rocky valleys and scrub desert. A common endemic to the western Cape and Namibia. *Call.* A scolding 'chrrrr-chrrrr' and a short warbling song. *Afrikaans.* Namakwasuikerbekkie.

787 Whitebellied Sunbird *Nectarinia talatala* L 11 cm
The only sunbird in the region to have a bottle-green head and breast, and a white belly. Female is very similar to the female Yellowbellied Sunbird but shows no trace of yellow on the white underparts. *Habitat.* Mixed moist and dry woodland, and suburbia. Common and widespread in the north and east. *Call.* A loud 'pichee, pichee', followed by a burst of fast, buzzy notes. *Afrikaans.* Witpenssuikerbekkie.

786 Yellowbellied Sunbird *Nectarinia venusta* L 11 cm
A bright yellow-bellied form of the Whitebellied Sunbird. It superficially resembles the Collared Sunbird but has a much longer, decurved bill, and is blue, not green on the back and head. Female resembles the female Whitebellied Sunbird but differs by having a pale yellow belly and flanks. *Habitat.* Miombo and broad-leafed riverine woodlands. Uncommon and localized in Zimbabwe and Moçambique, with vagrants further south. *Call.* A 'tsui-tse-tse' and a trilling song. *Afrikaans.* Geelpenssuikerbekkie.

793 Collared Sunbird *Anthreptes collaris* L 10 cm
Smaller than the similarly coloured Yellowbellied Sunbird but has a much shorter bill and a green, not blue throat extending only to the upper breast. The female, which resembles a small warbler, differs from the female Bluethroated Sunbird by having metallic green, not olive upperparts. *Habitat.* Evergreen, coastal and riverine forests, and mixed, moist woodland. Common in the extreme north, and in the east and south-east. *Call.* A soft 'tswee' and a harsher, chirpy song. *Afrikaans.* Kortsuikerbekkie.

794 Bluethroated Sunbird *Anthreptes reichenowi* L 10 cm
A small green sunbird which behaves like a warbler. Male shows a triangular blue-black throat patch and is unmistakable. The female's dull olive, not bright metallic green upperparts distinguish her from the female Collared Sunbird. *Habitat.* Mixed moist woodland and coastal forest. Common but localized in Moçambique. *Call.* A sharp 'tik-tik'. *Afrikaans.* Bloukeelsuikerbekkie.

795 Violetbacked Sunbird *Anthreptes longuemarei* L 13 cm
The male's violet head, back and tail are diagnostic and are conspicuous in direct sunlight. The female is a drab brown and white and only at close range can the violet-coloured tail be seen. Imm. lacks the violet on the tail and resembles an imm. Whitebellied or Yellowbellied Sunbird but has a very much shorter bill. *Habitat.* Miombo and moist mopane woodland. Uncommon and localized in Zimbabwe and Moçambique. *Call.* A sharp 'chit-chit' or 'skee'. *Afrikaans.* Blousuikerbekkie.

794 ▲

788 ▲

786 ▼   795 ▲

787 ▲   793 ▼

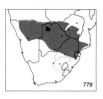

779 Marico Sunbird *Nectarinia mariquensis* L 14 cm
Distinguished from the very similar Purplebanded Sunbird by its larger size, noticeably longer and thicker bill, and broader purple breast band. Differs from the doublecollared sunbirds by its black, not grey belly and from Neergaard's and Shelley's sunbirds by its purple, not red breast band. Female is distinguished from female Purplebanded by being larger and having a longer, thicker bill. *Habitat.* Thornveld and dry broad-leafed woodland. Common in the north. *Call.* Typical chippering sunbird calls and a fast, warbling song. *Afrikaans.* Maricosuikerbekkie.

780 Purplebanded Sunbird *Nectarinia bifasciata* L 12 cm
Smaller than the similarly coloured Marico Sunbird, this species also has a thinner, shorter and less decurved bill. The female is less yellow below and has a shorter bill than the female Marico Sunbird. The male differs from the male Neergaard's Sunbird by having a purple, not red breast band. *Habitat.* Thornveld, moist broad-leafed woodland and coastal scrub. Common in the north-east and east. *Call.* A high-pitched 'teeet-teeet-tit-tit' song, accelerating at the end. *Afrikaans.* Purperbandsuikerbekkie.

782 Neergaard's Sunbird *Nectarinia neergaardi* L 10 cm
Within its restricted range, this green and black sunbird cannot be confused with any other sunbird. It is a small bird, with a thin, short, decurved bill, a black belly and a narrow red breast band. The female differs from the female Purplebanded Sunbird (with which it associates) by its shorter bill and much paler, unstreaked underparts. *Habitat.* Thornveld and sandy broad-leafed woodland. A common but very localized endemic to Zululand and southern Moçambique. *Call.* A thin, wispy 'weesi-weesi-weesi' and a short, chippy song. *Afrikaans.* Bloukruissuikerbekkie (Neergaardse Suikerbekkie).

783 Lesser Doublecollared Sunbird *Nectarinia chalybea* L 12 cm
Smaller than the Greater Doublecollared Sunbird but most easily distinguished from that species by having a shorter bill and a narrower red breast band. While the females also differ in bill character, this species is generally greyer below. In comparison with the female Orangebreasted Sunbird, the female Lesser Doublecollared is smaller and has greyish underparts. Range does not overlap with Neergaard's and Miombo Doublecollared sunbirds. *Habitat.* Coastal scrub, fynbos and forest edge, generally in less mountainous terrain than the Greater Doublecollared. Common and widespread endemic to the south and east. *Call.* A harsh 'chee-chee' and a short, fast song. *Afrikaans.* Klein-rooiborssuikerbekkie.

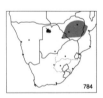

784 Miombo Doublecollared Sunbird *Nectarinia manoensis* L 13 cm
Occurring in the same habitats as the very similar Shelley's Sunbird, this species is differentiated by its pale olive, not black belly. In the field, it is virtually indistinguishable from the Lesser Doublecollared Sunbird unless this species' grey rump is seen; however, their ranges are exclusive so confusion is unlikely. *Habitat.* Miombo woodland and the edges of montane forest. Common in Zimbabwe, Moçambique and north-eastern Botswana. *Call.* The same as that of the Lesser Doublecollared Sunbird. *Afrikaans.* Miombo-rooiborssuikerbekkie.

785 Greater Doublecollared Sunbird *Nectarinia afra* L 14 cm
Far larger than the Lesser Doublecollared Sunbird, this species has a longer, heavier bill and a much broader red breast band. Females of both species are dark brown and olive but this one is greener below and has a longer, heavier bill. *Habitat.* More mountainous terrain than Lesser Doublecollared Sunbird. Common in the south, east and central regions. *Call.* A harsh, frequently repeated 'tchut-tchut-tchut' and a fast song. *Afrikaans.* Groot-rooiborssuikerbekkie.

782 ▲   779 ▼   785 ▼

780 ▼   783 ▼   784 ▼

801 House Sparrow *Passer domesticus* L 14 cm

The most familiar town and garden bird in the region, it is easily recognized by its grey crown and rump, reddish brown back, and black throat. Distinguished from the Great Sparrow by lacking the bright chestnut back and rump. Female and imm. are a dull grey-brown, and show a narrow white eye-stripe which differentiates them from the Yellowthroated Sparrow. *Habitat.* Towns, cities and gardens; rarely absent from human habitation. Common throughout the region. *Call.* Various chirps, chips and a 'chissick'. *Afrikaans.* Huismossie.

803 Cape Sparrow *Passer melanurus* L 15 cm

The male is unmistakable: it is the only sparrow in the region to have a pied head. Female is distinguished from the female House and Great sparrows by having a grey head with faint white shadow markings of the male's pied head pattern, and a chestnut back. *Habitat.* Grasslands, grain fields, and near human habitation. A common endemic to the central, western and southern regions. *Call.* A sharper, clearer call than that of the House Sparrow. *Afrikaans.* Gewone Mossie.

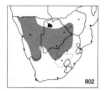

802 Great Sparrow *Passer motitensis* L 15 cm

Differs from the House Sparrow by its larger size, bright chestnut back and sides of head, and chestnut, not grey rump. Female is larger and much redder on the back than the female House Sparrow. *Habitat.* Dry thornveld; not usually associated with human habitation. Locally common in the dry central and western regions. *Call.* A liquid 'cheereep, cheereeu'. *Afrikaans.* Grootmossie.

804 Greyheaded Sparrow *Passer griseus* L 15 cm

The chestnut rump, back and wings, combined with the unmarked grey head, should preclude confusion with any other sparrow. Female House, Great and Cape sparrows superficially resemble this species but all have white or buff head markings. *Habitat.* Mixed woodlands and, in some regions, suburbia. Common and widespread but absent from the south-west. *Call.* Various chirping notes. *Afrikaans.* Gryskopmossie.

805 Yellowthroated Sparrow *Petronia superciliaris* L 15 cm

The dark head with its broad, creamy white eyebrow stripe is the best field character. The yellow breast spot is not often seen except at close range. Distinguished from the much larger Whitebrowed Sparrow-weaver by its lack of a white rump. *Habitat.* Thornveld, broad-leafed woodland, and riverine bush. Common in the north-eastern, central and eastern regions. *Call.* A three-part, whistled 'trrreep-trrreep-trrreep'. *Afrikaans.* Geelvlekmossie.

801 ▲   802 ▼   804 ▼   805 ▲

803 ▲

798 **Redbilled Buffalo Weaver** *Bubalornis niger* L 24 cm
The robust red bill, black plumage and white wing patches are diagnostic in this large weaver. Female and imm. are brown versions of the black male. In flight, might be mistaken for a Palewinged Starling but the chunky bill should eliminate confusion. *Habitat.* Dry thornveld and broad-leafed woodland. A nomadic weaver, common in the north and north-west. *Call.* Song is a 'chip-chip-doodley-doodley-dooo'. *Afrikaans.* Buffelwewer.

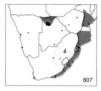

807 **Thickbilled Weaver** *Amblyospiza albifrons* L 18 cm
The only dark weaver to have a massive, thick bill, white wing patches and a small white spot at the base of the bill. Female and imm. are heavily streaked below but still show the diagnostic heavy bill. *Habitat.* Reedbeds, and coastal and evergreen forests. Common in northern Botswana, and in the east and south-east. *Call.* A 'tweek, tweek' flight call, and chattering when breeding. *Afrikaans.* Dikbekwewer.

808 **Forest Weaver** *Ploceus bicolor* L 16 cm
In the region, the only weaver with black upperparts and bright yellow underparts. Female and young have dark brown upperparts, grizzled forehead and throat, and yellow underparts. *Habitat.* Evergreen, riverine and coastal forests. Common in the east and south-east. *Call.* Song is a 'cooee-cooee-squizzzz', and variations of this. *Afrikaans.* Bosmusikant.

799 **Whitebrowed Sparrow-weaver** *Plocepasser mahali* L 19 cm
This large, plump, short-tailed weaver is distinctive with its broad white eyebrow stripe and conspicuous white rump. The smaller Yellowthroated Sparrow has a buff, not white eyebrow stripe and lacks the white rump. Birds in the north show faint speckling across the breast. *Habitat.* Thornveld and dry, scrubby rivercourses. Common in the dry central and western regions. *Call.* A harsh 'chik-chik' and a loud, liquid 'cheeoop-preeoo-chop' whistle. *Afrikaans.* Koringvoël.

800 **Sociable Weaver** *Philetairus socius* L 14 cm
The black chin, black barred flanks and scaly-patterned back render this species unmistakable. They build huge communal nests which appear to 'thatch' the trees in which they are built. When not frequenting their nests, small flocks roam the veld. *Habitat.* Dry thornveld and broad-leafed woodland. Common endemic to the north-west. *Call.* A chattering 'chicker-chicker' flight call. *Afrikaans.* Versamelvoël (Familievoël).

302

798 ▲

799 ▲

800 ▼   807 ▲

808 ▼

814 Masked Weaver *Ploceus velatus* L 15 cm
The breeding male is distinguished from the Lesser Masked Weaver by its brown legs, red, not white eye, and by the black mask, which does not extend behind the eye, forming a point on the throat. Differs from the larger Spottedbacked Weaver by having a uniform green, not yellow-spotted back. *Habitat.* Breeds colonially in trees overhanging water, but sometimes nests far from water in thornveld and suburbia. Common and widely distributed throughout the region except in the south-east. *Call.* A sharp 'zik, zik' and the usual swizzling weaver notes. *Afrikaans.* Swartkeelgeelvink.

815 Lesser Masked Weaver *Ploceus intermedius* L 15 cm
In breeding plumage, the black mask helps distinguish this species from the very similar Masked Weaver: it extends well on to the crown and comes to a rounded, not pointed end on the throat. This species also shows a white, not red eye and has blue, not brown legs. *Habitat.* Breeds colonially in trees overhanging rivers and dams, and in reeds. Locally common in the north and east. *Call.* The typical swizzling sounds of weavers. *Afrikaans.* Kleingeelvink.

811 Spottedbacked Weaver *Ploceus cucullatus* L 17 cm
Male is the only black-faced weaver in the region to have a yellow-spotted back. Distinguished from Masked Weaver by having a yellow crown. The northern race has a black crown and is then differentiated from the Masked Weaver by the yellow-spotted back. *Habitat.* Colonial nester in reedbeds, trees overhanging water and, in thornveld, sometimes away from water. Forms large flocks and frequents grasslands when not breeding. Common in the north, east and south-east. *Call.* A throaty 'chuk-chuk' and buzzy, swizzling notes. *Afrikaans.* Bontrugwewer.

812 Chestnut Weaver *Ploceus rubiginosus* L 15 cm
The black head, chestnut back and underparts render the breeding male unmistakable. Females and non-breeding males are drab, sparrow-like birds with grey-brown, rather than greenish coloration, and a well-defined yellowish throat which ends in a brownish breast band. *Habitat.* Thornveld and dry, broad-leafed woodland. Locally common only in Namibia. *Call.* Usual 'chuk, chuk' and swizzling weaver-type notes. *Afrikaans.* Bruinwewer.

819 Redheaded Weaver *Anaplectes rubriceps* L 15 cm
The male's scarlet head, breast and mantle are diagnostic. Females and non-breeding males have lemon yellow heads, show diagnostic orange-red bills, and have clear white bellies and flanks. *Habitat.* Thornveld, and mopane and miombo woodlands. Locally common in the north-east. *Call.* 'Cherrra-cherrra' and a high-pitched swizzling. *Afrikaans.* Rooikopwewer.

817 Yellow Weaver *Ploceus subaureus* L 15 cm
Smaller and a brighter yellow than the Golden Weaver, this species has a more compact bill and a red, not yellow eye. It is distinguished from the larger, pale-eyed Cape Weaver by having a less noticeable orange wash across the face, and by having more vivid yellow plumage. *Habitat.* Breeds in reedbeds, and trees near water, but is usually found in coastal and riverine forests. Common in the east and south-east. *Call.* Softer 'chuks' and swizzling than other, larger weavers. *Afrikaans.* Geelwewer.

819▲   814▼   811▼   817▲

815▼   812▼

813 Cape Weaver *Ploceus capensis* L 17 cm
Distinguished from the smaller Yellow Weaver by its heavier bill, less brilliant yellow plumage and by the orange wash over its face. Differs from both the Brownthroated and Golden weavers by being greener on the back, and by having an orange wash over the face and forehead. It lacks the well-defined chestnut bib of the Brownthroated Weaver from which it further differs by having a pale, not dark eye. *Habitat.* Builds communal nests in reedbeds and trees. A common endemic to the south-western and central regions. *Call.* A harsh 'azwit, azwit' and swizzling noises. *Afrikaans.* Kaapse Wewer.

818 Brownthroated Weaver *Ploceus xanthopterus* L 15 cm
The brightest of all the 'yellow' weavers, the breeding male with its chestnut bib is unmistakable. Distinguished from the Cape Weaver by being a more brilliant yellow, by lacking the orange wash over the face and crown, and by having a brown, not pale eye. *Habitat.* Breeds over water in reedbeds and forages in adjoining forest and scrub. Uncommon and localized in the east and in northern Botswana. *Call.* A soft 'zweek, zweek' and swizzling notes. *Afrikaans.* Bruinkeelwewer.

810 Spectacled Weaver *Ploceus ocularis* L 15 cm
Diagnostic features of both sexes are the sharp, pointed bill, the black line through the very pale yellow eye, and the orange head. Male has a black bib. A shy, skulking weaver which is more often heard than seen. *Habitat.* Coastal, evergreen and riverine forests, and moister areas in thornveld. Common in the eastern sector of the region. *Call.* Song is a descending 'dee-dee-dee-dee-dee'. *Afrikaans.* Brilwewer.

809 Oliveheaded Weaver *Ploceus olivaceiceps* L 14 cm
In the male the plumage combination is unmistakable: golden crown, olive cheeks and orange breast. Female has an olive head and back, and bright yellow underparts. *Habitat.* Miombo woodland. Locally common in Moçambique. *Call.* Not recorded in our region. *Afrikaans.* Olyfkopwewer.

816 Golden Weaver *Ploceus xanthops* L 18 cm
Distinguished from the smaller Yellow Weaver by its large, heavy, black bill and yellow, not red eye. Male in breeding plumage becomes a duller yellow and shows a green back. *Habitat.* Breeds in reedbeds, riverine scrub, and trees. Uncommon in the south-east, becoming locally more common in the extreme north, north-east and east. *Call.* A typical weaver-like 'chuk' and swizzling calls. *Afrikaans.* Goudwewer.

809▲

813▲ 810▼

816▼ 818▶

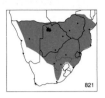

821 Redbilled Quelea *Quelea quelea* L 13 cm
Breeding males are unmistakable small finches with black faces and bright red bills and legs. Non-breeding females and males are drab, sparrow-like birds, which also show red bills and legs. The breeding female has a brown bill. Often encountered in huge flocks. *Habitat.* Dry, mixed woodlands, farmlands and grasslands. Common to extremely abundant in the greater part of the region, but absent from the south and dry west. *Call.* Flocks make a chittering noise. Song is a jumbled mixture of harsh and melodious notes. *Afrikaans.* Rooibekkwelea.

822 Redheaded Quelea *Quelea erythrops* L 12 cm
Male differs from male Redbilled Quelea by having a greater expanse of red on the head and by lacking the black face and red bill. The female Redheaded Quelea has a much more yellow wash across the face and breast than the Redbilled Quelea. *Habitat.* Damp grasslands and adjoining woodland. Locally common and nomadic in the north-east. *Call.* Described as a soft twittering. *Afrikaans.* Rooikopkwelea.

823 Cardinal Quelea *Quelea cardinalis* L 13 cm   □
Doubtfully recorded within the region. *Afrikaans.* Kardinaalkwelea.

824 Red Bishop *Euplectes orix* L 14 cm
Breeding male very similar to Firecrowned Bishop but differs by having a black forehead and crown, and by lacking black primaries (even in non-breeding plumage). Difficult to distinguish the female and non-breeding male from similar plumage stage of the Golden Bishop but this species has darker, more heavily streaked underparts. *Habitat.* Breeds in reedbeds, but at other times large flocks frequent grasslands. Common and widespread but absent from the central and north-western regions. *Call.* In display, males give a buzzing, chirping song. Normal flight call is a 'cheet-cheet'. *Afrikaans.* Rooivink.

825 Firecrowned Bishop (Blackwinged Bishop) *Euplectes hordeaceus*
    L 14 cm
Distinguished from Red Bishop in breeding plumage by having a darker mantle, black primaries and a red, not black forehead and crown. In flight, females and non-breeding males are distinguished from the similarly plumaged Red Bishop by showing black primaries. *Habitat.* Damp grassy areas and reedbeds in miombo woodland. Uncommon and localized in the north-east. *Call.* Described as being similar to that of the Red Bishop. *Afrikaans.* Vuurkopvink.

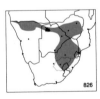

826 Golden Bishop *Euplectes afer* L 12 cm
Breeding male is distinctive with its black and yellow plumage. Difficult to distinguish from Red Bishop when not breeding, but is generally much paler below with faint streaking confined to the sides of the breast and the flanks. *Habitat.* Breeds in reedbeds surrounding dams, and near rivers. When not breeding, nomadic in large flocks over grasslands. Common in the central, northern and north-western regions. *Call.* Buzzing and chirping notes, similar to those of Red Bishop. *Afrikaans.* Goudgeelvink.

821▲   826▲   824▼ 825▼   822▲

829 Whitewinged Widow *Euplectes albonotatus* L 15-19 cm
The only widow in the region to have white on the primary coverts. Breeding male distinguished from Yellowbacked Widow by having a shorter tail, no yellow on its back and by having white in the wings. Easily identified in non-breeding plumage by its yellow shoulders and white patch at the base of the primaries. *Habitat.* Breeds in damp, grassy areas. When not breeding, flocks frequent grassland and thornveld. Common but localized in the east, and in the central, eastern and northern regions. *Call.* A nasal 'zeh-zeh-zeh' and a repetitive 'witz-witz-witz'. *Afrikaans.* Witvlerkflap.

827 Yellowrumped Widow (Cape Bishop) *Euplectes capensis* L 15 cm
In comparison with the Yellowbacked Widow, this species has a much shorter tail, and a yellow rump and lower back – not a yellow mantle. Female differs from the female Yellowbacked Widow by being far more heavily streaked below, and by having a dull yellow rump. *Habitat.* Damp grassy areas, bracken-covered mountain slopes and valleys, and fynbos. Common and widespread in the south and east. *Call.* Male calls a 'zeet, zeet, zeet' and a harsh 'zzzzzzzzt' in flight. *Afrikaans.* Kaapse Flap.

830 Yellowbacked Widow *Euplectes macrourus* L 15-22 cm
In the region, the only widow with a long black tail and a yellow back. Distinguished from the Yellowrumped Widow by its longer tail and lack of yellow on the lower back and rump, and from the Whitewinged Widow by the lack of white in the wings. Female is less heavily streaked on the breast and lacks the dull yellow rump of the female Yellowrumped Widow. She also differs from the female Whitewinged Widow by lacking any white or yellow in the wings. *Habitat.* Damp grasslands and marshy areas. Locally common in Zimbabwe. *Call.* Described as a buzzing sound. *Afrikaans.* Geelrugflap.

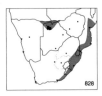

828 Redshouldered Widow *Euplectes axillaris* L 19 cm
The only small, short-tailed widow with a red shoulder. Distinguished from non-breeding and female Longtailed Widows by its much smaller size, less rounded, floppy wings and lack of buff stripe below the red shoulder. *Habitat.* Reedbeds, damp grasslands and stands of sugar-cane. Common and widespread in the east and the central northern regions. *Call.* Various twittering and chirping sounds are given by the male during display. *Afrikaans.* Kortstertflap.

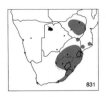

831 Redcollared Widow *Euplectes ardens*
L 15 cm (plus 25 cm tail)
The breeding male is unmistakable with its long wispy tail and its black plumage offset by a red crescentic throat patch. Female, imm. and non-breeding males are difficult to distinguish from the similarly coloured and sized bishop and widow birds, but they do show a bold, black- and buff-striped head pattern and have unstreaked buffy underparts. *Habitat.* Grassy and bracken-covered mountain slopes and, frequently, in sugar-cane in Natal. Common in scattered localities in the eastern sector. *Call.* A fast, high-pitched 'tee-tee-tee-tee-tee' is given by displaying males. *Afrikaans.* Rooikeelflap.

832 Longtailed Widow *Euplectes progne* L ♂ 19 cm (plus 40 cm tail); ♀ 16 cm
The largest widow of the region. The breeding male is unmistakable with its extra-long tail, and bright red shoulder bordered by a buff stripe. Male in non-breeding plumage is distinguished from the Redshouldered Widow by its larger size, broad, rounded wings and buff stripe below the red shoulder. The female is far larger than all other similarly coloured widows. *Habitat.* Open grasslands, especially in valleys and damp areas. Common in the central and south-eastern regions. *Call.* The male's call is a 'cheet, cheet' and a harsher 'zzit, zzit'. *Afrikaans.* Langstertflap.

829▲ 827▼

828▼ 830▲

832▼ 831▲

860 **Pintailed Whydah** *Vidua macroura* L 12 cm (plus 22 cm tail)
Breeding male differs from male Shaft-tailed Whydah by being white, not buff below and by lacking spatulate ends to the tail. Female and non-breeding male are difficult to distinguish from the similarly plumaged Shaft-tailed Whydah, although this species' bright red bill and more boldly striped black and white head aid identification. *Habitat.* Frequents a variety of grassland areas. Common throughout the region except the dry north-west. *Call.* Displaying males give 'tseet-tseet-tseet' calls. *Afrikaans.* Koningrooibekkie (Koningweeduweetjie).

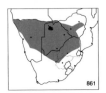

861 **Shaft-tailed Whydah** *Vidua regia* L 12 cm (plus 22 cm tail)
Breeding male has buff and black plumage, not black and white as in Pintailed Whydah, and has diagnostic spatulate tips to its tail. The female and non-breeding male have less distinct head markings than the similarly plumaged Pintailed Whydah. Flocks of small whydahs seen in northern Namibia are of this species. *Habitat.* Grassy areas in dry thornveld and broad-leafed woodland. Common and widespread in the dry central and western regions, eastwards to southern Moçambique. *Call.* Similar to but harsher than that of the Pintailed Whydah. *Afrikaans.* Pylstertrooibekkie (Pylstertweeduweetjie).

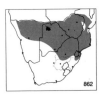

862 **Paradise Whydah** *Vidua paradisaea* L 15 cm (plus 23 cm tail)
Breeding male differs from the Broadtailed Paradise Whydah by having a longer tail which tapers to a point. In the field, the non-breeding male, female and imm. are indistinguishable from the similarly plumaged Broadtailed Paradise Whydah. *Habitat.* Mixed woodland, especially thornveld. Common in the northern sector of the region. *Call.* A sharp 'chip-chip' and a short 'cheroop-cherrup' song. *Afrikaans.* Gewone Paradysvink.

863 **Broadtailed Paradise Whydah** *Vidua obtusa* L 15 cm (plus 15 cm tail)
Identical in coloration to the breeding male Paradise Whydah, the breeding male of this species has a shorter tail, which is broad and rounded at the tip. In the field, the non-breeding male, female and imm. are indistinguishable from the similarly plumaged Paradise Whydah. *Habitat.* Miombo woodland, thornveld and riverine scrub. Locally common in the extreme north-east. *Call.* Described as being very similar to that of the Paradise Whydah. *Afrikaans.* Breëstertparadysvink.

860 ▲

862 ▼ 863 ▼

861 ▲

864 Black Widowfinch *Vidua funerea* L 11 cm
Breeding male is distinguished from breeding male Purple and Steelblue widowfinches by the combination of a greyish white bill and red legs and feet. In the field, the non-breeding male, female and imm. are indistinguishable from similarly plumaged widowfinches. *Habitat.* Forest edge, thornveld and riverine scrub. Common in the eastern sector of the region. *Call.* A short, canary-like jingle, and a scolding 'chit-chit-chit'.
Afrikaans. Gewone Blouvinkie.

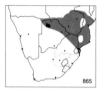

865 Purple Widowfinch *Vidua purpurascens* L 11 cm
In breeding plumage the combination of white bill, and whitish or pale pink legs and feet distinguish both male and female from other widowfinches. Indistinguishable from other widowfinches when in non-breeding and imm. plumages. *Habitat.* Thornveld and dry, broad-leafed woodland. Locally common in the south of its range, becoming more common in the north-east. *Call.* Described as mimicking the calls and song of Jameson's Firefinch. *Afrikaans.* Witpootblouvinkie.

866 Violet Widowfinch *Vidua incognita*   ☐
A newly described species which parasitizes the Brown Firefinch in the extreme north of our region. Very little is known about this widowfinch and details of its field identification features are required.
Afrikaans. Persblouvinkie.

867 Steelblue Widowfinch *Vidua chalybeata* L 11 cm
In breeding plumage, the combination of red bill, legs and feet distinguish both male and female from other widowfinches. The non-breeding male and female and the imm. are indistinguishable from similarly plumaged widowfinches. *Habitat.* Common and widely distributed in the north and north-east. *Call.* A canary-like song which includes clear, whistled 'wheeet-wheeet-wheeetoo' notes. *Afrikaans.* Staalblouvinkie.

820 Cuckoo Finch *Anomalospiza imberbis* L 13 cm
Similar to a small female 'yellow' weaver but its yellow colouring is brighter, especially below, and it has a diagnostic short, heavy, black bill. The imm. is more buffy and, although its bill is not black, it still shows the short, stubby character. *Habitat.* Open, upland grasslands, especially near damp areas. Uncommon and localized in the central, northern and eastern areas. *Call.* Described as a 'tsileu, tsileu'. A swizzling noise is uttered during display. *Afrikaans.* Koekoekvink.

314

864▲ 867▼ 820▼ 865▲

840 Bluebilled Firefinch *Lagonosticta rubricata* L 10 cm
The distinct grey, not pink crown and nape differentiate this species from the very similar Jameson's Firefinch. In the northern race, the whole head is washed pinkish and the Bluebilled is then distinguished from Jameson's Firefinch by showing no trace of pink on its darker back and wings. *Habitat.* Thickets in thornveld, riverine scrub and forest edges. Common in the south-east, north to Moçambique. *Call.* A fast, clicking 'trrt-trrt-trrt-trrt', like a fishing reel as the line is played out, and a 'wink-wink-wink'. *Afrikaans.* Kaapse Robbin.

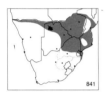

841 Jameson's Firefinch *Lagonosticta rhodopareia* L 10 cm
The entire head, back and wings of this species are bright pink: it does not show the dark back and wings of the Bluebilled Firefinch. Female is far brighter than the female Bluebilled Firefinch and has a pinkish, not brown back and wings. *Habitat.* Thickets and grassy tangles in dry thornveld. Locally common in the north and north-east. *Call.* A clicking trill similar to that of the Bluebilled Firefinch, but higher pitched and interspersed with a sharp 'vit-vit-vit'. *Afrikaans.* Jamesonse Robbin.

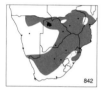

842 Redbilled Firefinch *Lagonosticta senegala* L 10 cm
Although similar to the Brown Firefinch, the male has more extensive pink on the face, throat and breast, which is only faintly speckled with white. It has a pink, not dark rump; a feature which distinguishes the female from the female Brown Firefinch. *Habitat.* Mixed woodland, especially near water, and throughout suburbia in Zimbabwe. Common, but absent from the dry west and the southern regions. *Call.* A sharp, fast 'vut-vut-vut-chit-chit' and a 'sweeep'. *Afrikaans.* Rooibekrobbin.

843 Brown Firefinch *Lagonosticta nitidula* L 10 cm
Easily distinguished from the female Redbilled Firefinch by its diagnostic dark, not pink rump, and from the male Redbilled Firefinch by having far less pink on the face, throat and breast, which is densely speckled with white. *Habitat.* Found in thick scrub and reeds close to water. Locally common in the extreme northern and central regions. *Call.* Described as a guttural trill, and a 'tsiep, tsiep' flight call. *Afrikaans.* Bruinrobbin.

806 Scalyfeathered Finch *Sporopipes squamifrons* L 10 cm
An unmistakable small finch with a black and white freckled forehead, black malar stripes, and white-fringed wing and tail feathers. Imm. like ad. but lacks the freckling on the forehead and the black malar stripes. *Habitat.* Dry thornveld, cattle kraals and frequently around farm buildings. A common endemic to the dry central and western regions. *Call.* A soft 'chizz, chizz, chizz', given by small groups when in flight. *Afrikaans.* Baardmannetjie.

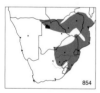

854 Orangebreasted Waxbill *Sporaeginthus subflavus* L 10 cm
Superficially resembles the swees but it lacks the grey on the head and has a bright yellow-orange belly and barred flanks. Distinguished from the Quail Finch (which frequents the same habitat) by having a red, not dark rump, and by its call. *Habitat.* Grasslands and weedy areas, especially near water. Common, sometimes in large flocks, in the north and east. *Call.* A soft, clinking 'zink zink zink' flight call. *Afrikaans.* Rooiassie.

840 ▲

841 ▲

842 ▲ 806 ▼ 854 ▼ 843 ▲

857 Bronze Mannikin *Spermestes cucullatus* L 9 cm
Distinguished from the Redbacked Mannikin by having less black on the head and by having a brown, not chestnut back. Differs from the much larger Pied Mannikin by its less massive, bi-coloured bill. Imm. is a uniform dun brown, unlike the imm. Redbacked Mannikin which has a reddish brown back. *Habitat.* Frequents a wide variety of grassy areas in woodlands, forest edges and damp regions. Common in the eastern sector. *Call.* A soft, buzzy 'chizza, chizza'. *Afrikaans.* Gewone Fret.

858 Redbacked Mannikin *Spermestes bicolor* L 9 cm
The conspicuous chestnut back, greater amount of black on the head and throat, and the pale grey bill distinguish this species from the Bronze and Pied mannikins. Imm. differs from imm. Bronze Mannikin by its chestnut-tinged back. *Habitat.* Moist, broad-leafed woodland and coastal forest. Common in the eastern sector of the region. *Call.* A thin, soft 'seeet-seeet', uttered when flushed from grass. *Afrikaans.* Rooirugfret.

859 Pied Mannikin *Spermestes fringilloides* L 13 cm
The largest mannikin of the region, it is easily distinguished from the Bronze and Redbacked mannikins by its markedly black and white appearance, its large, heavy bill, and the chestnut bars on its flanks. Imm. is paler below than imm. Redbacked Mannikin and lacks the chestnut wash over its back. *Habitat.* Bamboo stands, and riverine and coastal forests. Uncommon and very localized in the north-east and east. *Call.* Described as a chirruping 'peeoo-peeoo'. *Afrikaans.* Dikbekfret.

855 Cut-throat Finch *Amadina fasciata* L 12 cm
The pinkish red band across the throat is diagnostic in the male. Female is smaller than the female Redheaded Finch, is more boldly barred on the head, and has the back distinctly streaked and mottled. *Habitat.* Dry thornveld and broad-leafed woodland. Locally common in the north. *Call.* An 'eee-eee-eee' flight call. *Afrikaans.* Bandkeelvink.

856 Redheaded Finch *Amadina erythrocephala* L 13 cm
Distinguished from the Redheaded Quelea (p. 308) by having heavily barred underparts. Female differs from the smaller female Cut-throat Finch by having the head uniformly brown, and the under- and upperparts less heavily barred and streaked. *Habitat.* Dry grasslands, thornveld and broad-leafed woodland. A common endemic to the drier western regions. *Call.* A soft 'chuk-chuk', and a 'zree, zree' flight call. *Afrikaans.* Rooikopvink.

858 ▲

859 ▲

855 ▲   856 ▼   857 ▲

844 Blue Waxbill *Uraeginthus angolensis* L 13 cm
Its powder blue face and breast render this species unlikely to be confused
with any other finch. The female is a paler, 'faded' version of the male.
Habitat. The drier areas of mixed woodland, and suburbia in some regions.
Common in the north and north-east. *Call.* A soft 'kway-kway-sree-seee-seee-
seee'. *Afrikaans.* Gewone Blousysie.

845 Violeteared Waxbill *Uraeginthus granatinus* L 15 cm
Male is unmistakable. Female is biscuit-coloured, like the Blue Waxbill, but
can be distinguished by the dull violet ear patches, the blue, not dark rump,
and by the lack of any blue on the underparts. *Habitat.* Dry thornveld,
rivercourses and grassy road verges. Common and widely distributed in the
drier western regions, east to Moçambique. *Call.* A soft, whistled 'tiu-woowee'.
Afrikaans. Koningblousysie.

846 Common Waxbill *Estrilda astrild* L 13 cm
This small, brownish, long-tailed finch is best distinguished by its red bill and
face patch, and by the small reddish patch on the belly. Imm. lacks the red bill
and face patch but does show the red belly patch. *Habitat.* Long grass in
damp areas, alongside rivers and in reedbeds. Common throughout the
region except the dry central areas. *Call.* A nasal 'cher-cher-cher' and a 'ping,
ping' flight note. *Afrikaans.* Rooibeksysie (Rooibekkie).

847 Blackcheeked Waxbill *Estrilda erythronotos* L 13 cm
The greyish brown body and head, conspicuous black face, and dark red
rump and flanks, render this species unmistakable. Female and imm. are
duller versions of the male. *Habitat.* Grassy areas and thick tangles in dry
thornveld. Common in the dry central and western regions. *Call.* A high-
pitched 'chuloweee'. *Afrikaans.* Swartwangsysie.

848 Grey Waxbill *Estrilda perreini* L 11 cm
Could be confused with the Cinderella Waxbill; however, their distributions are
mutually exclusive, and this species differs by having the red confined to the
rump and uppertail coverts. Although they are similar and also show the red
rump, the East African Swee and the Swee Waxbill lack the grey on the body
and have green backs. *Habitat.* Edges of evergreen forests, and in thick
coastal and riverine forests. Locally common in the east and south-east.
Call. The waxbill's typical, soft 'pseeu, pseeu'. *Afrikaans.* Gryssysie.

849 Cinderella Waxbill *Estrilda thomensis* L 11 cm
The western counterpart of the Grey Waxbill, it is distinguished by being
generally paler grey, having a red base to the bill, and having the red on the
rump extending to the lower belly and flanks. *Habitat.* Riverine scrub and
forest. A species of southern Angola which enters our region only in the
extreme north of Namibia. *Call.* Not recorded in our region.
Afrikaans. Swartoogsysie.

845 ▲

847 ▲

844 ▼   846 ▲

848 ▲   849 ▼

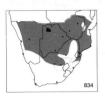

834 Melba Finch *Pytilia melba* L 12 cm

A much more brightly coloured finch than the Goldenbacked Pytilia, the male shows a crimson face, bill and throat, blue-grey nape, boldly barred belly and flanks, and it lacks the orange wing panel. Female differs from female Goldenbacked Pytilia by being brighter, more boldly marked below, and by lacking the orange wing panels. *Habitat.* Thornveld and dry, broad-leafed woodland. Common in the northern sector of the region. *Call.* A short 'wick, wick', and a song consisting of a whistled 'trrreeee-chrrroooo'. *Afrikaans.* Gewone Melba (Melbasysie).

833 Goldenbacked Pytilia *Pytilia afra* L 11 cm

The male could be confused with the male Melba Finch, but it lacks blue-grey on the nape, is less distinctly barred on belly and flanks, has a dull red face, and shows an orange wing panel. Female is distinguished from the female Melba Finch by being drabber, less distinctly barred below and by showing the orange wing panel. *Habitat.* Thick, tangled scrub, usually near water and often in miombo woodland. Uncommon and localized in the north-east. *Call.* Described as a 'seee', and a two-note, piping whistle. *Afrikaans.* Geelrugmelba (Geelrugsysie).

850 Swee Waxbill *Estrilda melanotis* L 10 cm

The grey head, black face and red rump are diagnostic in the male. The female lacks the black face and is, like the imm., indistinguishable in the field from both sexes of the East African Swee. However, confusion is unlikely as their distribution ranges do not overlap. *Habitat.* Long grass and scrub at forest edges, usually in mountainous areas. A common but localized endemic to the southern and eastern areas. *Call.* A soft 'swee-swee' call is uttered in flight. *Afrikaans.* Suidelike Swie (Suidelike Swie-sysie).

851 East African Swee *Estrilda quartinia* L 10 cm

Lacks the black face of the male Swee Waxbill and is indistinguishable from the female of that species; however, their ranges are mutually exclusive. Superficially resembles the Orangebreasted Waxbill but the bright yellow-orange belly and the barred flanks of that species are absent. *Habitat.* Forest edges in mountainous terrain. Locally common in the highlands of Zimbabwe and Moçambique. *Call.* The same as that of the Swee Waxbill. *Afrikaans.* Tropiese Swie (Tropiese Swie-sysie).

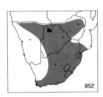

852 Quail Finch *Ortygospiza atricollis* L 9 cm

Rarely seen on the ground but, when flushed from the grass, it has a diagnostic buzzing flight and a distinctive call. When seen, the black and white face pattern, barred black and white breast and flanks, and the lack of red on the rump, distinguish it from the Orangebreasted Waxbill. *Habitat.* Open grasslands. Common and widespread throughout the region except in the dry west. *Call.* In flight, gives a continual, tinny 'chillink, chillink' call. *Afrikaans.* Gewone Kwartelvinkie.

853 Locust Finch *Ortygospiza locustella* L 9 cm

Although similar to Quail Finch in habits, their calls are very different. Male is easily told by its red face, throat, and breast, rust-coloured wings, and black back and belly. Female is distinguished from the female Quail Finch by her rust-coloured wings. *Habitat.* Grasslands, especially in damper areas. Uncommon and localized in Zimbabwe and Moçambique. *Call.* A fast 'tinka-tinka-tinka' given in flight. *Afrikaans.* Rooivlerkkwartelvinkie.

322

834 ▲

853 ▲

851 ▲

852 ▼

833 ▼

850 ▲

838 Pinkthroated Twinspot *Hypargos margaritatus* L 12 cm
The male is much paler than the male Redthroated Twinspot and has a pinkish face and breast. Female is distinguished from the female Redthroated Twinspot by showing no trace of red or pink on the grey-brown throat and breast. *Habitat.* Thicker tangles of thornveld and coastal scrub. Locally common endemic to southern Moçambique and northern Natal. *Call.* A soft, reedy trill. *Afrikaans.* Rooskeelrobbin.

839 Redthroated Twinspot *Hypargos niveoguttatus* L 12 cm
Distinguished from Pinkthroated Twinspot by its deep red, not pinkish face, throat and breast. Female differs from female Pinkthroated Twinspot by showing an orange-red wash across the breast. *Habitat.* Prefers more moist situations than the Pinkthroated Twinspot, and the edges of evergreen forests. Uncommon and localized endemic to the east. *Call.* A trill similar to that of the Pinkthroated Twinspot. *Afrikaans.* Rooikeelrobbin.

836 Redfaced Crimsonwing *Cryptospiza reichenovii* L 12 cm
Superficially resembles the Nyasa Seedcracker but has a black, not red tail, a dull red, not dark green back and wings and, on the head, the red is confined to eye patches. Female lacks the male's red eye patches and differs from the female Nyasa Seedcracker by having red in her wings. *Habitat.* The understorey and thick tangles of evergreen forests. Uncommon and localized in the highlands of Zimbabwe and Moçambique. *Call.* Described as a 'zeet'. *Afrikaans.* Rooirugrobbin.

837 Nyasa Seedcracker *Pyrenestes minor* L 13 cm
Differs from the Redfaced Crimsonwing by having the red on the face extending on to the throat, by having a red, not black tail, and by lacking any red in the wings. Female closely resembles the male but has a reduced red face patch. *Habitat.* Thickets along small streams, and near water in forest clearings and miombo woodland. Uncommon to locally common in Zimbabwe and Moçambique. *Call.* Described as a 'tzeet' and a sharp, clipped 'quap'. *Afrikaans.* Rooistertrobbin.

835 Green Twinspot *Mandingoa nitidula* L 10 cm
Smaller than the Redfaced Crimsonwing and the Nyasa Seedcracker, it differs from both by having its black belly and flanks spotted with white. Unlikely to be confused with the Redthroated or Pinkthroated twinspots because neither of those species shows green in their plumage. *Habitat.* Thickets in evergreen, coastal and riverine forests. Although easily overlooked because of its secretive habits, it is locally common in the east and north-east. *Call.* A soft, rolling, insect-like 'zrrreet'. *Afrikaans.* Groenrobbin.

838▲

839▼

837▼

836▲

835▼

872 Cape Canary *Serinus canicollis* L 13 cm
Distinguished from the Yelloweyed Canary by lacking bold facial markings, having a yellow forehead and crown, and by the greater amount of grey on the nape encompassing the sides of the neck. Imm. differs from the female Yellow Canary by having its greenish yellow underparts overlaid with heavy brown streaking. *Habitat.* Prefers mountainous terrain. Common and widespread in the south, east, and sporadically further north. *Call.* A clear 'sklereee', and in flight, a twittering tinkle. *Afrikaans.* Kaapse Kanarie.

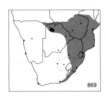

869 Yelloweyed Canary *Serinus mozambicus* L 12 cm ˈ
Where their ranges overlap, this species could be confused with the Yellow Canary. It has a distinctive black stripe through the eye, and a moustachial stripe combined with a blue-grey nape, which Yellow Canary never shows. Distinguished from Bully Canary by its smaller size and less robust bill. Imm. has buffy yellow, lightly streaked underparts. *Habitat.* Thornveld, mixed woodland and suburbia. Common to abundant in the east. *Call.* A 'zeee-zereee-chereeo'. *Afrikaans.* Geeloogkanarie.

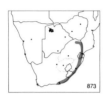

873 Forest Canary *Serinus scotops* L 13 cm
A very dark, heavily streaked canary which has a small black chin and a yellow eye-stripe. Imm. is duller and lacks the black chin. *Habitat.* Evergreen forests, edges and clearings. Common but localized endemic to the south-east and east. *Call.* A soft 'tsisk', and a quiet, jumbled song. *Afrikaans.* Gestreepte Kanarie.

874 Cape Siskin *Serinus tottus* L 12 cm
Closely resembles Drakensberg Siskin but the male shows white tips to its secondary and tail feathers. The ranges of the two species are mutually exclusive. *Habitat.* Montane fynbos, exotic pine plantations, and sometimes along the coast. Common but localized endemic to the extreme south. *Call.* Flight call is described as a 'pitchee'. *Afrikaans.* Kaapse Pietjiekanarie.

875 Drakensberg Siskin *Serinus symonsi* L 12 cm
Resembles only the Cape Siskin, and their ranges never overlap. Male lacks the white-tipped secondary and tail feathers of the Cape Siskin. Imm. is very similar to female but is more heavily streaked below. *Habitat.* Scrub-filled valleys and montane grasslands. Common endemic to higher altitudes of the Drakensberg. *Call.* Similar to the Cape Siskin. *Afrikaans.* Bergpietjiekanarie.

877 Bully Canary *Serinus sulphuratus* L 15 cm
The darker, southern birds can be distinguished from Yelloweyed and Yellow canaries by larger size, a chunkier bill and greenish wash across the breast. The paler, smaller northern birds differ from Yellow Canary by having a more massive bill and by lacking the contrasting yellow rump. *Habitat.* Frequents a wide range of scrub, woodlands and grasslands, but prefers coastal scrub. Common along the coast from the western Cape north to Moçambique, and into the Transvaal and Zimbabwe. *Call.* Deeper pitched and slower than other canaries. *Afrikaans.* Dikbekkanarie.

878 Yellow Canary *Serinus flaviventris* L 13 cm
The dark form resembles Bully and Yelloweyed canaries but it lacks both the greenish breast of the former, and the black eye-stripe and moustachial streak, and grey nape of the latter. Pale form differs from pale form Bully Canary by being a deeper, brighter yellow, and by showing a contrasting yellow rump and a less massive bill. Female differs from imm. Cape and Yelloweyed canaries by having white underparts streaked with brown. *Habitat.* Karoo and coastal scrub, and scrubby mountain valleys. Common endemic to the drier central and western regions. *Call.* A fast, jumbled series of 'chissick' and 'cheree' notes. *Afrikaans.* Geelkanarie.

326

872 ▲

875 ▲

869 ▲

873 ▲

874 ▼

878 ▼

877 ▲

870 **Blackthroated Canary** *Serinus atrogularis* L 11 cm
A dull, brown and grey canary with a diagnostic black-speckled throat, and a bright yellow rump which contrasts with the otherwise nondescript body colour. Distinguished from the larger Whitethroated Canary, which has a greenish yellow rump, by its less robust bill and black-speckled throat. *Habitat.* Thornveld, dry broad-leafed woodland, and near waterholes in dry regions. Common in the central and dry western regions. *Call.* A prolonged series of wheezy whistles and chirrups. *Afrikaans.* Bergkanarie.

871 **Lemonbreasted Canary** *Serinus citrinipectus* L 10 cm
Superficially similar to but smaller than the Yelloweyed Canary with which it sometimes flocks. Male has a pale lemon throat and breast, well demarcated from the remainder of its buff underparts. It lacks the bold head markings and grey nape of Yelloweyed Canary, and has a small white ear patch. Female resembles a very buff Blackthroated Canary but lacks the black-speckled throat. *Habitat.* Thornveld, open grasslands and palm savanna. A locally common endemic to Moçambique and northern Natal. *Call.* Similar to that of the Blackthroated Canary, but higher pitched and shorter. *Afrikaans.* Geelborskanarie.

876 **Blackheaded Canary** *Serinus alario* L 13 cm
The male has a bright chestnut back and tail, a black head, and black stripes down the sides of the breast. The black head is variable and can appear pied. The female lacks the black markings but does show the diagnostic chestnut back and resembles a small female Cape Sparrow. *Habitat.* Fynbos, Karoo scrub and mountainous terrain. A locally common endemic to the south-west and west. *Call.* A soft 'sweea' or 'tweet'. *Afrikaans.* Swartkopkanarie.

879 **Whitethroated Canary** *Serinus albogularis* L 15 cm
Differentiated from other dull-coloured canaries by the combination of its small white throat patch and greenish yellow rump. In similar habitat, distinguishable from female Yellow Canary by being larger, and by lacking breast streaking. *Habitat.* Fynbos, Karoo scrub and scrub-filled mountain valleys. Common endemic to the dry central and western areas. *Call.* A mixture of canary- and sparrow-like notes. *Afrikaans.* Witkeelkanarie.

880 **Protea Canary** *Serinus leucopterus* L 15 cm
A dark canary distinguishable from Whitethroated Canary by lacking the greenish yellow rump and by its small black chin. The narrow white edgings to its secondary coverts are not obvious but, if seen, are diagnostic. *Habitat.* Thick, tangled valley scrub and dense stands of proteas on hillsides. Locally common endemic to the extreme south. *Call.* Contact call is a 'tree-lee-loo'. Song is loud and varied. *Afrikaans.* Witvlerkkanarie.

881 **Streakyheaded Canary** *Serinus gularis* L 15 cm
Distinguished from Whitethroated Canary by having a finely streaked black and white crown, bordered by a broad white eyebrow stripe, and by lacking both the obvious white throat patch and the greenish yellow rump. Differs from Blackeared Canary by lacking the streaked breast and distinct black ear patches. *Habitat.* Various mixed woodland and scrub areas. Common in the south and east but absent from most of Moçambique. *Call.* A soft, weak 'trrreet', and a short song. *Afrikaans.* Streepkopkanarie.

882 **Blackeared Canary** *Serinus mennelli* L 14 cm
Distinguished from Streakyheaded Canary with which it sometimes occurs by the very distinct black cheeks and streaked breast. Female and imm. lack the black cheeks of the male but still show a streaked breast. *Habitat.* Mostly in miombo woodland. Common but localized in the north-east. *Call.* Described as a short 'teeu-twee-teeu, twiddy-twee-twee'. *Afrikaans.* Swartoorkanarie.

328

879▲

882▲

880▲

870▲   876▼ 881▼   871▲

868 Chaffinch *Fringilla coelebs* L 15 cm
The male is the only finch of the region to have a pinkish face and breast, a blue-grey head, and conspicuous white wing bars. Female is a dowdy grey-brown, sparrow-like bird but still shows clear white wing bars. *Habitat.* Exotic pine and oak plantations, and well-wooded gardens. Uncommon and localized on the western slopes of the Cape Peninsula. *Call.* A clear 'pink, pink, pink' and a short, rattling song. *Afrikaans.* Gryskoppie.

883 Cabanis's Bunting *Emberiza cabanisi* L 15 cm
Distinguished from the Goldenbreasted Bunting by having a greyish, not chestnut mantle, and black cheeks without a white stripe below the eye. Female and imm. are duller versions of the male. *Habitat.* Miombo woodland. Uncommon and localized in the north-east. *Call.* A clear 'tsseeoo' contact note. Song described as a 'wee-chidder-chidder-wee'. *Afrikaans.* Geelstreepkoppie.

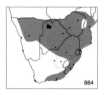

884 Goldenbreasted Bunting *Emberiza flaviventris* L 16 cm
Although very similar to Cabanis's Bunting, this species differs by having a white stripe below the eye, a chestnut, not greyish mantle, and a richer yellow breast, washed with orange. Female and imm. are duller versions of the male. *Habitat.* Thornveld, broad-leafed woodland and exotic plantations. Common and widespread throughout except for the southern, central and mid-western regions. *Call.* A nasal, buzzy 'zzhrrrr'. Song is a varied 'weechee, weechee, weechee'. *Afrikaans.* Rooirugstreepkoppie.

885 Cape Bunting *Emberiza capensis* L 16 cm
The greyish breast, chestnut wing coverts, and the lack of black on the throat distinguish this species from the Rock Bunting. Imm. has less definite head markings and duller chestnut wings. *Habitat.* Sandy coastal scrub in the south, and mountainous terrain further north. Common and widespread, but absent from the central, northern, and eastern coastal regions. *Call.* A buzzy, ascending 'zzoo-zeh-zee-zee', and a short chirping song. *Afrikaans.* Rooivlerkstreepkoppie.

886 Rock Bunting *Emberiza tahapisi* L 14 cm
In comparison with the Cape Bunting, this species has cinnamon, not grey underparts, a black, not white throat, and it lacks the chestnut wing coverts. Female and imm. have less bold black and white head markings but still show the diagnostic cinnamon underparts. *Habitat.* Rocky slopes in mountainous terrain, and mixed woodland in the north. Common in scattered localities in the northern and eastern sectors of the region. *Call.* A grating, rattled song and a soft 'pee-wee'. *Afrikaans.* Klipstreepkoppie.

887 Larklike Bunting *Emberiza impetuani* L 14 cm
A dowdy, nondescript bird which looks like a lark but behaves in typical bunting fashion, hopping over stones and grubbing for seeds on bare ground. Lacks any diagnostic field characters but the pale cinnamon wash over the breast might aid identification. *Habitat.* Dry rocky valleys and open plains, thornveld and dry broad-leafed woodland. While it is a common endemic to the drier western sector, it is nomadic in the east. *Call.* A soft, nasal 'tuc-tuc' and a short, snappy song. *Afrikaans.* Vaalstreepkoppie.

868 ▲

884 ▼   885 ▼   883 ▲

886 ▼   887 ▼

SOUTHERN OCEAN BIRDS

A **Emperor Penguin** *Aptenodytes forsteri* L 112 cm
Distinguished from the smaller King Penguin (p. 32) by its longer, noticeably decurved bill, very pale yellow ear patches, and by having a greater amount of black on the head extend over the nape. Imm. like ad. but has buff, not yellow ear patches. *Habitat.* Not found beyond the limits of the pack ice. Locally common on ice floes and bay ice within the Antarctic. *Call.* Normally silent but does give a nasal trumpeting.

B **Adelie Penguin** *Pygoscelis adeliae* L 71 cm
The short stubby bill and the black head with white eye-rings, render the ad. unmistakable. Imm. differs from Chinstrap Penguin by lacking the thin black line across the throat and by having the black on the head extend below the eye. *Habitat.* Rocky headlands and islands, pack ice and bay ice within the Antarctic. Common to abundant. *Call.* A fast, drumming 'ku-ku-ku' and a guttural 'gu-gu-gu-gu'.

C **Chinstrap Penguin** *Pygoscelis antarctica* L 76 cm
Ad. is unmistakable with its diagnostic thin black line across the throat. Ad. distinguished from imm. Adelie Penguin by having the black on the head not extending below the eye and, at close range, the black throat stripe can be seen. *Habitat.* Rocky headlands, islands, pack and bay ice. Common summer breeder on Bouvet Island. Vagrant to Marion and Gough islands. *Call.* Similar to that of Adelie Penguin but slower and deeper in pitch.

D **Gentoo Penguin** *Pygoscelis papua* L 76 cm
The triangular white ear patches and white-flecked head are diagnostic. The bill and feet are bright orange. Imm. is duller than ad. and has a grey, mottled throat. When seen at sea, this species is easily identified by its white ear patches. *Habitat.* Turf and gently sloping grassy areas near beaches. A common, all-year-round breeder in small colonies on Marion and Prince Edward islands. *Call.* A trumpeting 'ah aha aha aha-e'.

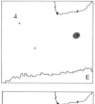

E **Imperial Cormorant** *Phalacrocorax atriceps* L 61 cm
A large and obvious black and white cormorant which has a bright blue eye, and which shows yellow at the base of its bill when breeding. The only cormorant on the Prince Edward islands. *Habitat.* Cliffs and rocky outcrops over the sea. Common resident in small colonies only on Marion Island. *Call.* Various croaks and hisses.

905 **Laysan Albatross** *Diomedea immutabilis* L 80 cm
Could be confused with an imm. Blackbrowed Albatross (p. 34) but differs by having a paler underwing which shows a variable amount of black streaking in the centre, especially towards the inner secondaries. *Habitat.* A species of the north Pacific which has been seen once off the southern Cape. *Call.* Not recorded in this region.

E▲

905▲

A ▲

B▼ D▼

▲ C

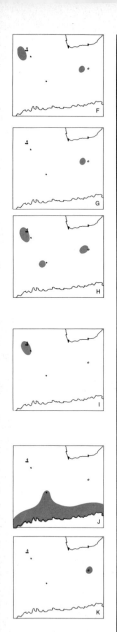

F **Common Diving Petrel** *Pelecanoides urinatrix* L 20 cm
In the field, indistinguishable from the Georgian Diving Petrel and, when in the hand, differentiated only by bill shape and structure. In flight, resembles a tiny puffin-like bird with its fast, whirring wingbeats. *Habitat.* Loose colonial breeders on turf and grassy ledges on cliffs and steep banks. A common summer breeder on Prince Edward Island and the Tristan group. *Call.* A low, moaning 'whoooaa whoooaa', given nocturnally from burrows.

G **Georgian Diving Petrel** *Pelecanoides georgicus* L 19 cm
Indistinguishable from the Common Diving Petrel except when in the hand, where bill shape and structure can be examined. *Habitat.* A common summer breeder on the Prince Edward group. Burrows are usually found in unvegetated laval scree and cinder cones. *Call.* Similar to that of the Common Diving Petrel.

H **Greybacked Storm Petrel** *Garrodia nereis* L 17 cm
Distinguished from the larger Whitebellied Storm Petrel (p. 46) by lacking the white rump. When seen from above, it might be confused with the Whitefaced Storm Petrel which also has a grey rump, but the completely dark head and breast of this species should prove distinctive. Flight is very swallow-like and the bird dances and patters over the water, often skipping from side to side. *Habitat.* Rare offshore at Prince Edward, Marion, Bouvet, and the Tristan group. Breeding not proven on any of these islands. *Call.* Described as a high-pitched twittering.

I **Whitefaced Storm Petrel** *Pelagodroma marina* L 21 cm
The distinctive, bold white face with a clear dark stripe through the eye, differentiates the Whitefaced from the smaller Greybacked Storm Petrel, even though both show a grey rump. When seen from above, the grey rump of this species contrasts slightly with its darker back and tail, and its feet project well beyond the tail. Flight is wild and erratic, with the bird often bounding and swinging buoyantly in different directions. *Habitat.* Common offshore at Gough Island and, on summer nights, is frequently attracted to the lights of the research station. *Call.* A soft, moaning 'woo woo woo'.

J **Snow Petrel** *Pagodroma nivea* L 34 cm
An unmistakable, medium-sized, pure white petrel with dark eyes, bill and legs. Away from pack ice, an albino of another similar-sized petrel could be mistaken for this species. Found in small flocks, often in the company of Antarctic Petrels, over open leads in pack ice. *Habitat.* Seldom far from pack ice. Possibly breeds on Bouvet Island. *Call.* A scolding 'tec, tec, tec'.

K **Kerguelen Tern** *Sterna virgata* L 31 cm
In all plumages, this species is distinguished from the Antarctic Tern (p. 144) by its much darker appearance, and its thinner, shorter and weaker bill. In breeding plumage shows a more obvious white cheek stripe, and grey underwing and tail. Imm. differs from imm. Antarctic Tern by being more heavily mottled above, by having a complete brown cap, and underparts heavily suffused with buff and brown. *Habitat.* Over the inshore kelp zone, inland grassy plains and freshwater lakes. Rare on the Prince Edward group. *Call.* A harsh 'chittick' and a drawn-out 'keeeaar'.

L **Lesser Sheathbill** *Chionis minor* L 38 cm
An unmistakable plump and pigeon-like white bird with a black bill and face, and pink eye-ring, legs and feet. Highly inquisitive, and confident in the presence of man. *Habitat.* Coastal grassy plains and on the edges of penguin colonies. Common on the Prince Edward group. *Call.* Soft clucks as they strut around, and a 'chak chak' in flight.

334

J▶

F▲

I▲

G▲

H▼ L▼

K▲

M▲

N▲

O▲

Q▼ R▼

P▲

M **Gough Moorhen** *Gallinula nesiotes* L 27 cm
The only rail on Gough Island, it resembles the Moorhen (p. 104) but differs by
having thick red legs, reduced white striping on its flanks, and very short,
rounded wings. *Habitat.* Thick tangles of bracken, tussock grass and *Phylica*
scrub on coastal plateaux. Common on Gough Island and recently introduced
to Tristan Island. *Call.* A sharp 'krrik'.

N **Inaccessible Island Rail** *Atlantisia rogersi* L 17 cm
A small black rail which scuffs, rodent-like, at the bases of tussock grass.
Frequently ventures out into the open to forage but scrambles back to cover if
alarmed. Imm. has a brown wash over the body. *Habitat.* Long tussock grass
and boggy pools near the coast. Common but confined to Inaccessible Island
in the Tristan group. *Call.* A soft, but carrying 'pseep'.

O **Tristan Bunting** *Nesospiza acunhae* L 16 cm
Occurs in association with Wilkins' Bunting and is easily distinguished by its
smaller, thinner bill, duller green plumage and lack of streaking on the
underparts. Found foraging in pairs or small family groups, it quickly dives for
cover if alarmed. *Habitat.* Confined to Inaccessible and Nightingale islands,
where it can be found virtually anywhere except on the windswept highlands.
Call. Not yet described.

P **Gough Bunting** *Rowettia goughensis* L 18 cm
The only passerine found on Gough Island. Female and imm. are buff, heavily
streaked with brown; the male is unusual in that it may be seen breeding and
heard singing in this 'immature' plumage. Ad. male is a uniform, drab olive
green with a small black bib and yellow eye markings. *Habitat.* An endemic to
Gough Island, it is most commonly found in thick tussock and briar tangles
adjoining rivers, but can be seen virtually anywhere on the island. *Call.* Song is
a rattling jingle. Usual call note is a sharp 'keet-keet'.

Q **Wilkins' Bunting** *Nesospiza wilkinsi* L 18 cm
Although the Wilkins' and Tristan buntings occur on the same islands,
confusion is unlikely to arise as Wilkins' Bunting has a massive, chunky bill,
specially adapted to crack open the hard *Phylica* nuts. It further differs from
the Tristan Bunting by having streaked flanks and brighter yellow underparts.
Habitat. Associated with *Phylica* scrub, bracken and tree fern tangles in
valleys on Inaccessible and Nightingale islands in the Tristan group.
Call. Not known.

R **Tristan Thrush** *Nesocichla eremita* L 22 cm
A very dark brown, heavily mottled thrush with rust-coloured wing patches,
especially noticeable in flight. Localized and furtive on Tristan Island, but
bolder and more confiding on the other islands. *Habitat. Phylica* scrub-filled
valleys, and bracken- and tree fern-covered slopes. Common on Tristan,
Inaccessible and Nightingale islands. *Call.* A soft 'seeep' and a short,
fluty song.

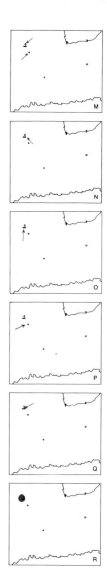

GLOSSARY

Accidental. A vagrant or stray species not normally found within the region.

Arboreal. Tree dwelling.

Axillaries. Group of feathers at the base of the underwing.

Carpal. The bend of the wing at the base of the primaries.

Cere. Bare, coloured skin at the base of a raptor's bill.

Colonial. Associating in close proximity, either while roosting, feeding or nesting.

Commensal. Living with or near another species, without being interdependent.

Coverts. Groups of feathers covering the bases of the major flight feathers or an area or structure such as the ear.

Crepuscular. Active at dawn and dusk.

Crest. Elongated feathers on the forehead, crown or nape.

Crown. The top of the head.

Cryptic. Pertaining to camouflage coloration.

Decurved. Curving downwards.

Diurnal. Active during daylight hours.

Eclipse plumage. Dull plumage attained by male ducks and sunbirds during a transitional moult after the breeding season and before they acquire brighter plumage.

Endemic. Restricted to a certain region. Southern Africa has over 140 endemic species.

Eyebrow. Usually referring to a stripe above the eye.

Eye-stripe. A stripe running from the base of the bill directly through the eye.

Feral. Species which have escaped from captivity and now live in the wild.

Field mark. A distinctive plumage mark which aids identification.

Flight feathers. The longest feathers on the wings and tail.

Flush. To rouse and put to flight.

Form. *See* Phase.

Frons. The forehead.

Frontal shield. Bare patch of skin at the base of the bill and on the forehead. Often brightly coloured.

Fulvous. Reddish yellow or tawny.

Gape. The basal opening of the bill.

Gorget. A colour patch on the throat.

Graduated tail. Refers to a pointed or rounded tail which assumes its shape in stages from the base.

Gular stripe. An expandable patch of bare skin on the throat.

Immature. A bird that has moulted from juvenile plumage but has not attained adult plumage. Can also include juvenile plumage.

Irruption. A rapid expansion of a species' normal range.

Juvenile. The first full feathered plumage of a young bird.

Local (movements). Short migrations.

Lore. The area between the base of the bill and the front of the eye.

Malar stripe. A stripe running downwards from the side of the lower mandible.

Mandible. A bird's beak or bill.

Mantle. The combined area of the back, upperwings and scapulars.

Migrant. A species which undertakes (usually) long-distance flights from its wintering to breeding areas.

Mirrors. The white spots on the primaries of gulls.

Montane. Pertaining to mountains.

Moustachial stripes. Stripes running from the base of the bill down the sides of the throat.

Nape. The upper hind neck and back of the head.

Nocturnal. Active at night.

Overwintering. A bird which remains in the sub-region instead of migrating to its breeding grounds.

Palearctic. North Africa, Greenland, Europe, and Asia north of the Himalayas, southern China and South East Asia.

Pelagic. Ocean dwelling.

Phase. A plumage colour stage.

Primaries. The outermost major flight feathers of the wing, nine, ten or twelve in number.

Race. A geographical population of a species; a subspecies.

Range. A bird's distribution.

Raptor. A bird of prey.

Recurved. Curving upwards.

Resident. A species not prone to migration, remaining in the same area all year.

Rictal bristles. Bare feather shafts at the base of the bill.

Rufous. Reddish brown.

Scapulars. Group of feathers on the back, at the base of the upperwing.

Secondaries. Longest wing feathers, from mid-wing to the base of the wing.

Shoulder. The area immediately in front of the carpals on a folded wing, or the carpal area.

Speculum. A patch of distinctive colour on the wing, especially in ducks.

Supercilium. Eyebrow stripe.

Territory. An area a bird establishes and then defends from others.

Tertials. The inner secondary feathers which lie close to the body.

Vagrant. Rare and accidental to the region.

Vent. The undertail region, extending to the legs.

SUGGESTED FURTHER READING

Berruti, A. and Sinclair, J.C. *Where to Watch Birds in Southern Africa.* Cape Town: C. Struik, 1983.

Harrison, P. *Seabirds: An Identification Guide.* London: Croom Helm, 1983.

Heinzel, H., Fitter, R., and Parslow, J. *The Birds of Britain and Europe with North Africa and the Middle East.* London: Collins, 1972.

King, B., Woodcock, M., and Dickinson, E.C. *A Field Guide to the Birds of South East Asia.* London: Collins, 1975.

Maclean, G.L. *Roberts' Birds of South Africa.* 5th ed. Cape Town: The Trustees of the John Voelcker Bird Book Fund, 1984.

Newman, K.B. *Birds of Southern Africa. I. Kruger National Park.* Johannesburg: Macmillan South Africa, 1980.

Newman, K.B. *Newman's Birds of Southern Africa.* Johannesburg: Macmillan South Africa, 1983.

Robbins, C.S., Bruun, B., and Zimm, H.S. *A Guide to the Field Identification: Birds of North America.* New York: Golden Press, 1966.

Sinclair, J.C., Mendelsohn, J.M., and Johnson, P. *Everyone's Guide to South African Birds.* Johannesburg: CNA, 1981.

Steyn, P. *Birds of Prey of Southern Africa.* Cape Town: David Philip, 1982.

SOCIETIES, CLUBS AND THEIR PUBLICATIONS

Most birders are members of their local or regional bird club. These clubs perform an important role in bringing birders together, both socially and through the medium of print. All clubs organize birding outings, usually in prime habitats, giving members the opportunity to meet and learn to bird-watch, regardless of their age. Most also arrange lectures and film shows, which serve to stimulate interest in and discussion about birds.

The following clubs are affiliated to the Southern African Ornithological Society (SAOS), which entitles those who join to membership of the SAOS, and to receive its two journals: the *Ostrich,* which is scientific in content, and *Bokmakierie,* a popular birding journal with high-quality photographs and general interest articles.

SAOS (Head Office), P O Box 87234, Houghton 2041.

SAOS-affiliated clubs:

Transvaal
Northern Transvaal Ornithological Society, P O Box 4158, Pretoria 0001. (Publishes *Laniarius.*)
Witwatersrand Bird Club, P O Box 72091, Parkview 2122. (Publishes *WBC News-sheet.*)

Cape
Cape Bird Club, P O Box 5022, Cape Town 8000. (Publishes *Promerops.*)
Eastern Cape Wild Bird Society, P O Box 1305, Port Elizabeth 6000. (Publishes *The Bee-eater.*)

Natal
Natal Bird Club, P O Box 4085, Durban 4000. (Publishes *Albatross* and *The Fret.*)

Orange Free State
Orange Free State Ornithological Society, P O Box 6614, Bloemfontein 9300.

Other southern African bird clubs include:
The Botswana Bird-Watching Club, P O Box 71, Gaborone, Botswana,
The Zimbabwe Ornithological Society, P O Box 8382, Causeway, Zimbabwe,
and the Ornithologische Arbeitsgruppe, SWA, P O Box 67, Windhoek, Namibia.

Those with more specific interests are catered for by such specialist bird study groups as:
The African Seabird Group, Percy FitzPatrick Institute of African Ornithology, University of Cape Town,
Rondebosch 7700. (Publishes *The Cormorant.*)
The Vulture Study Group, P O Box 4190, Johannesburg 2000. (Publishes *Vulture News.*)

The Wildlife Society of Southern Africa, P O Box 44189, Linden 2104, also promotes an interest in
birding activities.

PHOTOGRAPHIC CREDITS

INDEX TO COMMON NAMES

The following numbers refer to the page on which the species' account appears.

343

INDEX TO SCIENTIFIC NAMES

INDEX TO AFRIKAANS NAMES

NOTES

NOTES